PEST MANAGEMENT IN TRANSITION

Westview Replica Editions

This book is a Westview Replica Edition. The concept of Replica Editions is a response to the crisis in academic and informational publishing. Library budgets for books have been severely curtailed; economic pressures on the university presses and the few private publishing companies primarily interested in scholarly manuscripts have severely limited the capacity of the industry to properly serve the academic and research communities. Many manuscripts dealing with important subjects, often representing the highest level of scholarship, are today not economically viable publishing projects. Or, if they are accepted for publication, they are often subject to lead times ranging from one to three years. Scholars are understandably frustrated when they realize that their first-class research cannot be published within a reasonable time frame, if at all.

Westview Replica Editions are our practical solution to the problem. The concept is simple. We accept a manuscript in camera-ready form and move it immediately into the production process. The responsibility for textual and copy editing lies with the author or sponsoring organization. If necessary we will advise the author on proper preparation of footnotes and bibliography. We prefer that the manuscript be typed according to our specifications, though it may be acceptable as typed for a dissertation or prepared in some other clearly organized and readable way. The end result is a book produced by lithography and bound in hard covers. Initial edition sizes range from 500 to 800 copies, and a number of recent Replicas are already in second printings. We include among Westview Replica Editions only works of outstanding scholarly quality or of great informational value, and we will continue to exercise our usual editorial standards and quality control.

PEST MANAGEMENT IN TRANSITION:
With a Regional Focus on the Interior West

Pieter de Jong, Project Coordinator

This volume examines current pest control strategies, introduces new alternatives for pest control in the interior West, and documents successful integrated pest management programs from across the nation. The contributors include leaders in alternative pest control research, representatives of regional and federal agencies, grower organizations, industry, and environmental groups, and farmers and ranchers.

Pieter de Jong is on the staff of the Wright-Ingraham Institute (Colorado Springs, Colorado), a nonprofit educational organization concerned with the conservation, preservation, and use of human and natural resources.

Proceedings of a conference on
Pest Control Strategies for the Future

March 30-31, 1978
Denver, Colorado

U.S. Environmental Protection Agency
Dallas Miller, Project Coordinator

and

Wright-Ingraham Institute
Pieter de Jong, Project Coordinator

PEST MANAGEMENT IN TRANSITION

With a Regional Focus on the Interior West

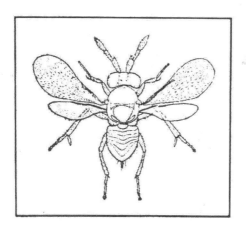

Pieter de Jong, Project Coordinator
WRIGHT-INGRAHAM INSTITUTE

Routledge
Taylor & Francis Group

LONDON AND NEW YORK

First published 1979 by Westview Press, Inc.

Published 2019 by Routledge
52 Vanderbilt Avenue, New York, NY 10017
2 Park Square, Milton Park, Abingdon, Oxon OX14 4RN

Routledge is an imprint of the Taylor & Francis Group, an informa business

Library of Congress Catalog Card Number: 79-53138

ISBN 13: 978-0-367-28279-0 (hbk)
ISBN 13: 978-0-367-29825-8 (pbk)

CONFERENCE CO-SPONSORS

Arkansas Valley Audubon Society, Pueblo
Denver Audubon Society
Denver Botanic Gardens
Enos Mills Group, Sierra Club, Denver
Environmental Protection Agency, Region VIII
Horticultural Advisory Council for El Paso County, Co.
Horticulture Arts Society of Colorado Springs
Ricon-Vitova Insectaries, Riverside, California
Rocky Mountain Farmers Union, Denver
Rodale Press, Inc., Emmaus, Pennsylvania

PARTICIPATING AGENCIES

Colorado Cooperative Extension Service
Colorado Department of Agriculture
Colorado State Forest Service
Forest Pest Management, Rocky Mountain Region (Forest Service, USDA)
Rocky Mountain Forest and Range Experiment Station
    (Forest Service, USDA)
Science Education Administration (USDA)
U.S. Department of Fish and Wildlife (Dept. of Interior)

CONFERENCE CONSULTANTS

Robert Simpson, Colorado State University
Beatrice Willard, Colorado School of Mines

PROCEEDINGS PRODUCTION STAFF

Pieter de Jong, manuscript preparation
John Torborg, graphics
Catherine Ingraham, typist

CONTENTS

Introduction
Insect Control
Weed Control
Urban Integrated Pest Management
Future Research Needs

PREFACE

Pest Management in Transition is the report of a two-day working conference entitled *Pest Control Strategies for the Future,* held 30, 31 March 1978 at the Botanic Gardens in Denver, Colorado. These proceedings include a diversity of perspectives offered by some of the nation's leaders in alternative pest control research, representatives of regional and federal agencies, grower organizations, industry, environmental groups, and farmers and ranchers.

The major aims of the conference were to examine current pest control strategies, introduce new alternatives for pest control into the Interior West[1], document successful integrated pest management programs from across the nation, and provide information on alternative pest control strategies to government agencies, educators, the agricultural community and concerned individuals.

The introduction of DDT, following World War II, signaled the beginning of the synthetic pesticide era. The initial success of pesticides led to a widespread reliance on pesticides. This increasing dependence on pesticides as the predominant pest control strategy has precipitated many negative effects, i.e. increased impact on non-target species and human health, reduction of naturally occurring biological controls, and increased resistance of pests to pesticides.

The publication of this proceedings comes during an important time of transition to integrated management. This report combines a regional view of pest problems and current control strategies in the Interior West with documentation of the economic and environmental soundness of integrated pest management programs from across the nation.

The efforts and cooperation of many individuals, agencies and organizations have contributed to these proceedings. The co-sponsors guaranteed a diverse and receptive audience and the participating agencies sent representatives who aided discussions on regional pest problems and control tactics. Kenneth Hood and Charles Reese of the Enviornmental Protection Agency in Washington, D.C. and Dallas Miller, EPA Region VIII, provided assistance in developing the program. Finally, the use of the Denver Botanic Gardens, made possible by William Gambill, was particularly appreciated as it provided a handsome and appropriate setting for the conference.

P.dJ.
Colorado Springs
September 1978

[1]Interior West defined as EPA Region VIII which includes Colorado, Utah, Montana, Wyoming, North and South Dakota.

INTRODUCTION

Elizabeth Wright Ingraham

On behalf of the Wright-Ingraham Institute and the Environmental
Protection Agency, Mr. Lehr and I would like to welcome you to this
Pest Control Strategies Conference. In introducing this conference,
I thought you might be interested in knowing how we got started on
the idea of integrated pest management. The Institute has been primarily interested in tackling ideas and issues at the interface of
humans and the environment. We have tried during the early years of
this new institution to look down the road to what may be important.

Last September, Pieter de Jong, a member of our administrative staff,
with whom most of you are familiar, came to the planning council of
the Institute and said he wanted to promote integrated pest management
in the region. We talked with a few people in agencies and organizations. One person retorted with, "Well, it's something of an esoteric
idea." At this point we wondered what was so esoteric about an idea
that affects, really deeply affects, food, timber and fiber production,
and an idea that involves both the rural and urban communities? One
difficulty, perhaps, was the newness of the integrated approach to
pest control, so we explored further and found the idea rising very fast
around the country. Our survey discovered, however, that no conferences
had yet been held in the Interior West. We gave the go-ahead to Pieter
who has put together what I think is an absolutely outstanding two-day
conference on this subject.

What all of us get out of it, how the proceedings come out and what
the final analysis is, cannot be predicted. In reviewing the development of any new idea, I think it's important to recognize that ideas are
a process. Many of us working on new ideas want them to move forward
immediately, but of course this doesn't often happen. New ideas have
to percolate through the system and find conduits for implementation.
We first have to expand the body of thought, which we are doing at this
conference. Then we have to take that thought and put it into demonstration and experimental models, prototypes and pilot projects. After
this the projects have to assessed and evaluated. That's a difficult
area because there the idea, as policy, must be reviewed by the political
forces before it becomes part of the system.

Although components of integrated pest management (IPM) are as old as the invention of agriculture, the concept of IPM is really very new. It was not until the mid-60's that the term was coined. In 1972, the Council on Environmental Quality put out its publication on IPM and while scientists had been working on the idea for many long years, it was now at the frontier of what we call the lead time for implementing an idea.

This important concept of IPM is closely tied and related, I think, to the rising interest in food production and environmental problems and the entire idea of food production to meet the demands of increased population. In 1975, when the Institute published a report on food production for the Kettering Foundation, this issue of pest management was raised. Pest control for crops was one of the key issues involved in considering global food production in the next 25 years. The issue is even more critical today. The Institute will continue to explore and expand on these integrated approaches which are necessary for a dynamic system. Out of this conference we hope to capture ideas on where we've been, where we're going, and what needs to be done to implement this new direction in pest control and management.

PART ONE
OVERVIEW

CURRENT PRACTICES
IN INSECT PEST CONTROL

David Pimentel and Nancy Goodman
Department of Entomology
Cornell University

INTRODUCTION

All crops and livestock are attacked by pests. Worldwide man loses
nearly half of his food to pests. World crop losses to pests (insects,
pathogens, weeds, mammals, and birds) are estimated to be about 35%
(Cramer, 1967). Mammal and bird losses appear to be more severe in
the tropics and subtropics than in the temperate region, but these
losses are low compared with losses to the three major pest groups of
insects, pathogens, and weeds.

In addition to the 35% preharvest loss an estimated 20% postharvest
loss results from another group of pests, primarily microorganisms,
insects, and rodents. When postharvest losses are added to preharvest
losses, worldwide food losses to pests are estimated to be about 48%
(35% preharvest plus 20% postharvest losses).

In the United States preharvest losses to pests are estimated to be
about 33% in spite of modern pest control technology (USDA 1965; Pimentel
(1976). This loss is not much below the estimate of the worldwide loss
of 35%. However, postharvest pest losses are about one half of the
worldwide level (20%) or only 9% (USDA 1965). Thus, total losses in the
United States are about 39%. This is a significant loss of valuable
food. As mentioned, these losses occur in spite of all pest management
efforts.

It is worthwhile to explain the term pest management and its relationship
to integrated pest management (IPM). Pest management is the general term
that includes all biological, cultural, and chemical programs employed
for pest control. Integrated pest management employs a combination of
biological and pesticidal controls. This is the way IPM was first defined and continues to be used today (R. Smith, University of California,
personal communication 1977).

The aim of this paper will be to examine the current use of biological,
cultural, and pesticidal controls in pest management in the United States.
In addition, we will briefly examine the environmental and social costs of
pesticide use.

BIOLOGICAL AND CULTURAL CONTROLS

Although pesticides are often considered to be the most important control technology for pests, biological and cultural controls in fact are more important than pesticides when a comparison is based on managed acres. For example, biological and cultural controls are employed on 9% of farm acreage compared with insecticidal controls that are employed on only 6% of the acres (Table 1). For the control of plant pathogens, some form of biological and cultural control is employed on more than 95% of the acreage compared with less than 1% on which fungicides are used. For weed control, the estimate is that nonchemical controls, primarily mechanical cultivation, are used on 80% of the acreage while only 17% are treated with herbicides (Table 1).

At this stage it would be profitable to examine some of the biological and cultural methods that are used to control pest insects, pathogens, and weeds.

Pests	Percentage of Acres Involved	
	Biological and Cultural Controls	Pesticidal Controls
Insects	9%	6%
Pathogens	90%	1%
Weeds	80%	17%

Table 1   Comparison of estimated biological and cultural and pesticidal controls employed on the United States Cropland for insects, pathogens and weeds (USDA, 1968; 1970; PSAC, 1965; Pimentel, 1976).

Host Plant and Animal Resistance

One of the most important reasons for serious pest problems on crops is the breeding of susceptible types (Lupton, 1977). When altering the genetic makeup of the crop plant to increase yields, in the past little or no attention was given to maintaining the natural resistance to pest attack that existed in the crop. Natural resistance can be lost or greatly reduced if care if not taken to maintain it. Of importance then is breeding plants that not only have high yields but are resistant to their major pests. Plant breeders have maintained the level of resistance to plant pathogens and are now giving vigorous attention to insect resistance in crops.

The differences in levels of resistance that may exist in crop plants and their effectiveness are well illustrated with pea aphids (*Acryrthosiphum pisum*) associated with alfalfa (*Medicago sativa*). Five young pea aphids placed on a common crop variety of alfalfa produced a total of 290 off-spring in ten days, whereas the same number of aphids for a similar period on a resistant variety of alfalfa produced a total of only two offspring per aphid (Dahms and Painter 1940). Obviously, a pest population with a 145-fold greater rate of increase on a host plant would inflict greater damage on the host plant than one with an extremely low rate of increase.

Sorghum provides another example. On a susceptible strain of commercial sorghum (*Sorghum vulgare*) the mean rate of oviposition (eggs per generation) of the chinch bug (*Blissus leucopterus*) was about 100. On a resistant strain of sorghum, however, the mean oviposition was less than one (Dahms 1948). In this instance, animal feeding was reduced by 99% on the resistant plants and had dramatic effects on the population dynamics of the feeding pests.

The Hessian fly (*Mayetiola destructor*), a serious pest of wheat, is effectively controlled on at least a third of Hessian fly infested acreage (20 million acres) by Hessian fly resistant varieties (PSCA 1965). Some biotypes of the Hessian fly have evolved that are able to overcome the resistance present in the wheat but new resistant varieties are being released.

Natural resistance to pests also exists in livestock. For example, European cattle introduced into South Africa were found to be more susceptible to the "bont" tick (*Amblyomma hebraeum*) and to "heartwater" disease than Afrikander (zebu) cattle (Bonsma 1944). The number of ticks on the European cattle were about four times more abundant than on the native Afrikander (table 2). The mortality in Afrikander cattle due to

	Totals for 12 cows of each kind		Ratio European:
	Afrikander	European	Afrikander
Number of ticks:			
12 counts per cow:			
On the body (800 cm^2)	237	1,775	7.5:1
On the escutcheon (200 cm^2)	1,529	4,397	2.9:1
Under the tail	2,140	4,789	2.2:1

Table 2 The number of ticks on Afrikander and European cattle (Bonsma 1944).

heartwater disease averaged only 5.3% over a thirty-month period whereas for European cattle the average was 60.7% mortality. Thus, the European cattle that had never been exposed to either the tick or heartwater disease were highly susceptible to the attack of both pests.

Rotations

In agriculture, when crops and livestock are maintained on the same land year after year, pests associated with the crops or livestock tend to increase in number and severity. For example, if cole crops are cultured for several years in the same soil, club root (*Plasmodiophora brassicae*) organisms increase rapidly and can totally ruin production (Walker et al. 1958). Planting noncole crops on the land for several years will effectively control these pests.

Also, if soybeans are grown on some land more than once every three to four years, brown stem rot (*Cephalosporius gregatum*) may be a serious problem (Carter and Hartwig 1963). In the U.S. corn belt, corn and small grains make good rotation crops and reduce problems from the brown stem rot in soybeans.

Rotating corn with soybeans also helps control the most serious insect pest of corn, corn rootworms (G. Musick and R. Treece 1975, personal communication). Corn must be rotated every year to provide effective control of rootworms, and this rotation combines nicely with soybeans and small grains. Corn rootworms are controlled successfully in about 60% of U.S. corn acreage by employing crop rotations (Pimentel et al. 1978a).

Cropping Systems

Although they may prefer one crop, some pests can feed on several crops. In addition, certain parasites and predators can attack a pest and related pests on different crops. Because the parasites and predators can move from one crop to another searching out suitable pest hosts, they can control pest populations. The technique requires planting suitable crops in combination. For example, plant bugs (*Lygus* spp.) feed on alfalfa and cotton. After alfalfa is mowed for hay and eliminated as a food source for the bugs, they will move to cotton in large numbers (Stern 1969). Thereafter, the bugs can damage cotton if present in sufficient numbers. Because the plant bugs generally prefer alfalfa, the Lygus bug population on cotton can be kept to a minimum by planned cutting of the alfalfa. The successful strategy is to cut only a portion of the alfalfa at a time leaving sufficient alfalfa to attract the bugs and keep them away from the cotton. Another strategy is to plant narrow strips of alfalfa (6 m wide) for every 91-122m of cotton in the cotton field. This not only attracts the plant bugs but may provide a source of natural enemies of such pests as cotton bollworms.

Also combining sorghum and cotton has demonstrated that several pests of cotton can be effectively reduced and the number of pesticide treatments significantly reduced (Ray Frisbie, Texas A&M; Don Peters, Oklahoma State Univ., personal communications 1975).

4

The leaf miner problem associated with spinach in California was in part solved by planting spinach considerably after the other crops that usually harbored the leaf miner (H. Lange, personal communication 1974). Without a suitable host, the leaf miner population was reduced significantly before the spinach was planted. Another way to control such a pest is to reduce the growing of crops that act as an alternate host.

Genetic Diversity

Numerous examples make it clear that many parasites have the genetic variability to evolve and overcome "single-factor" resistance in their host. Hence, although parasites associated with hosts in natural situations appear to be genetically stable, in agricultural ecosystems when stressed by a single factor, a parasite can often evolve, overcome host resistance, and cause serious damage to crops. For example, parasitic stem rust and crown rust have been found to overcome genetic resistance bred into their oat host. Since 1940, oat varieties have been changed in the corn belt region every four to five years to counter the changes in the races of stem rust and corn rust (Stevens and Scott 1950, van der Plank 1968).

Recently the Southern corn leaf blight parasite overcame resistance in corn (Thurston 1973). The use of Texas sources of cytoplasmically inherited male sterility (TMS) narrowed the resistance character in about 85% of the corn grown in the United States to almost genetic homogeneity (Moore 1970; Roane 1973). Then in 1970, favorable environmental conditions resulted in selecting race T of *Helminthosporium maydis* (Southern corn leaf blight), which is virulent on all plants with TMS cytoplasm. The resulting epidemic caused devastating losses in the genetically homogeneous corn host (Nelson et al. 1970).

Genetic diversity within a given crop, however, prevents a pest from overcoming natural plant resistance (Wolfe 1968). For example, when the wheat variety Eureka, resistant to wheat stem rust races, was grown in progressively larger acreages, the incidence of the rust races attacking Eureka also increased (Fig. 1). The prime reason for the increase was that with the greater distribution of the Eureka variety, rust races could now be more easily transmitted from host to host. When the abundance and distribution of Eureka wheat declined, the incidence of rust infections by the special rust race also declined (Fig. 1). This example of a pathogen-wheat host system clearly illustrates the benefits of genetic diversity in agricultural crops.

Pest outbreaks occurring in "green revolution" wheat and rice varieties have been associated with planting a single variety over wide regions (Frankel 1971; Ida Oka 1973, personal communication). The need for genetic diversity of resistant characters in host-plants for both "green revolution" and U.S. monocultures has been well documented by Pathak (1970), Adams et al. (1971), Smith (1971), and Day (1973).

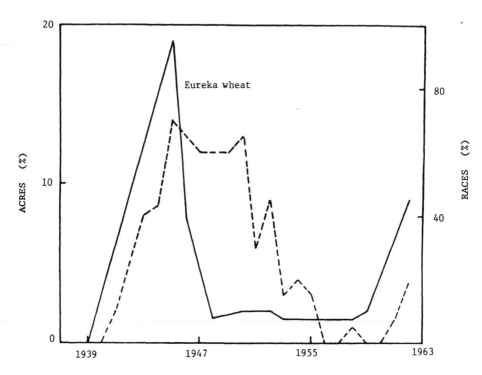

Figure 1   The percentage of acres of Eureka wheat (———) grown
in northern New South Wales of total wheat acres and
the percentage of wheat stem rust races which are able
to attack Eureka compared with all races present (----).
(Data of Watson and Lugig, 1963)

Planting Times

Some plants in nature begin to grow early or late in the growing season
and thereby manage to escape the attack of certain pests. Wild radishes
have been observed to germinate early in the spring and make most of their
growth before the cabbage maggot fly emerges and attacks the radishes.
Damage to the radishes under these conditions is usually minimal (Pimentel,
unpublished).

Other plants may escape attack by starting growth after the pest population has emerged and begun to die. This strategy has been employed by
agriculturalists to reduce the attack on wheat by the Hessian fly. De-
laying the seeding of wheat fields is effective on about 67% of the
wheat acreage in reducing the attack from the dangerous fall brood of
the Hessian fly (PSAC 1965). The spring brood of the Hessian fly is less
serious and delayed planting does not reduce the spring brood attack.

Growth stage may be an important factor in vulnerability to attack. Corn,
for example, is more susceptible to corn borer attack at certain stages
of growth. Young corn (15.2 cm high) is relatively unattractive to corn
borers compared with corn that has reached 45.7 cm high (Whitman 1975).
If a farmer wants to reduce corn borer attack, corn should be planted
so that plants are either very small or nearly full grown before the
borer moths emerge.

Introduced Natural Enemies

Biological control utilizing introduced parasites and predators has
proved highly effective in controlling certain insect pests. For example,
both the spotted alfalfa aphid and alfalfa weevil are major pests of
alfalfa. Of the 29 million acres of alfalfa grown in the United States,
about 9 million are infested with the spotted alfalfa aphid. Control of
this pest is achieved primarily by natural enemies and by the planting
of alfalfa varieties resistant to the aphid (PSAC 1965). Although nearly
half of the alfalfa crop is attacked by the alfalfa weevil, this pest
is now generally controlled by natural enemies and alfalfa culture
practices.

Citrus and olive crops on about 3 million acres have several insect pests
that are effectively controlled by natural enemies (Sweetman 1958, van
den Bosch and Messenger 1973). Control of a few of the citrus insect
pests is achieved on most of the citrus acreage whereas control of the
olive insect pests is generally effective on all acreage.

CHEMICAL PEST CONTROL

Various chemical methods including olfactory attractants, juvenile hormones, and especially pesticides are employed for control of insects,
pathogens, and weeds. The extent of the use of these controls is discussed.

Olfactory Attractants

Insect behavior is governed by various stimuli including chemicals released by some insect species themselves. These stimuli play important
roles in guiding insect feeding, mating, and oviposition.

Feeding, mating, and oviposition are all essential behavioral patterns
for insects. If the pattern is altered by changing the normal stimuli
that the animal receives, survival and reproduction may be significantly

reduced. The use of olfactory attractants is aimed at changing the normal stimuli received by insects to limit the increase of insect populations (Birch 1974; Roelofs 1975).

These attractants can be used to monitor insect populations by baiting traps to determine when to treat an insect population with an insecticide. Most agree that one of the extremely difficult tasks in economic entomology is to measure insect pest densities accurately to determine when to treat with an insecticide (PSAC 1965). The use of attractants for monitoring offers unique opportunities to improve the measurement of insect population densities.

Attractants have already been employed effectively to eradicate a serious pest, the Mediterranean fruit fly (*Ceratitus capitata*) from Florida in 1956 (Metcalf et al. 1962). In this case an attractant, protein hydrolysate bait containing malathion, was distributed throughout the area of fruit fly infestation (PSAC 1965). The use of the attractant made it possible to bring the flies directly in contact with an extremely small amount of the insecticide. This was the first actual eradication of a pest in the United States.

Chemical attractants can also be employed for baiting "sticky" traps to capture, for example, pest moths. On an experimental basis, Roelofs et al. (1970), controlled the redbanded leaf-roller (*Argyro-taenia velutinana*) from New York apple orchards by attaching a sticky trap to apple trees and baiting it with a female sex attractant. When about 100 traps per hectare were used, effective control was achieved without insecticide use.

Another technique employing a sex attractant that offers potential is the use of the attractant to "disrupt" normal mating of the pest population. For insect pests that totally depend upon a chemical sex attractant for mating, it may be possible to release sufficient quantities of the attractant to "disrupt" the mating of the population.

This procedure has several advantages. First, the technique is specific for a particular pest, avoiding the difficulties of the use of broad spectrum insecticides. Second, often the amount of chemical introduced into the environment is extremely small (as low as 3 grams per hectare per season)(W. Roelofs, Cornell University, personal communication 1976). Third, since all the sex attractants known are nontoxic to humans and most other life, the impact on the environment should be minimal.

Attempts to employ sex attractants against the introduced gypsy moth (*Porthetria dispar*) have had some success experimentally (Beroza et al. 1973). Application from 2 to 5 g per hectare of encapsulated pheromone (disparlure) reduced gypsy moth males by 90% or more for 6 to 8 weeks.

In another study, Shorey and co-workers (Birch 1974; Farkas et al. 1975) have experimentally demonstrated that releasing about 25 mg/ha/10-hr night of the pheromone, *cis*-7-dodecenyl acetate, could disrupt about 95%

of the cabbage looper (*Trichoplusia ni*) population. The actual amount of pheromone necessary for release to disrupt a population depends on the species and environmental conditions. For example, under windy conditions considerably more pheromone would have to be used.

The potential advantages of attractants have been mentioned, but the technique also has limitations. One of the most important limitations, which is also experienced with insecticides, is the evolution of "resistance" or "tolerance" in insect populations to attractants.

Although some insect species depend entirely upon sex attractants for mating, a "dosage response" does exist in their reaction to the attractant stimuli. Recall earlier that 90% of the gypsy moth population and 95% of the cabbage looper population were disrupted by the released sex attractant. The 5 to 10% of these populations that were not responding probably require a higher dosage or other stimuli. If some of the 5 to 10% are able to mate in spite of the disruption, then natural selection is operating and the populations should eventually evolve tolerance. In fact, populations of the cabbage looper have evolved tolerance and are able to carry out normal mating in spite of the release of a sex attractant (H.H. Shorey, University of California, Riverside, personal communication 1976).

Pesticide Use and Controls

The use of pesticides more than doubled during the 10 year period from 1966 to 1976; pesticide use on crops and livestock increased from 503 million pounds to over 1 billion pounds (Berry 1978).

Pesticide use in agriculture is not evenly distributed (Table 3). For example, 50% of all insecticide used in agriculture is applied to the nonfood crops of cotton and tobacco. Of the food crops, fruit and vegetables receive the largest amounts of insecticide. Of the herbicidal material applied, 45% is used on corn, with the remaining 55% distributed among numerous other crops (Table 3). Most of the fungicidal material is applied on fruit and vegetables, with only a small amount used on field crops (Table 3).

Benefits of Pesticides

Pesticides are essential to U.S. agricultural production; however, in spite of more than 1 billion pounds of pesticides used in agriculture, an estimated 33% of all crops is lost annually due to pest attack in the United States. This loss of food and fiber amounts to about $35 billion, or enough to pay for our 1976 oil imports.

We have examined the frequently asked question, what would our crop losses to pests be if all pesticides were withdrawn from use, and readily available nonchemical control methods were substituted where possible? It appears that crop losses based on dollar value would increase from the estimated 33% to about 41% (Pimentel et al 1978b). Thus we estimated an 8% increase in crop losses worth $8.7 billion could be anticipated if pesticides were withdrawn from use.

9

| Crops | Insecticides | | Herbicides | | Fungicides | | % of Total |
	%Acres	%Amount	%Acres	%Amount	%Acres	%Amount	Crop Acres
Nonfood	NA[1]	50	NA	NA	0.5	NA	NA
Cotton	61	47	82	9	4	1	1.11
Tobacco	77	3	7	NA	7	NA	0.11
Field Crops	NA	33	NA	NA	NA	15	NA
Corn	35	17	79	45	1	NA	7.43
Peanuts	87	4	92	2	85	11	0.16
Rice	35	1	95	3	0	NA	0.22
Wheat	7	1	41	5	0.2	NA	6.11
Soybeans	8	4	68	16	2	NA	4.19
Pasture Hay & Range	0.5	2	1	4	0	NA	68.40
Vegetable	NA	7	NA	2	NA	24	NA
Potatoes	77	2	51	NA	49	10	o.16
Fruit	69	9	NA	1	NA	60	NA
Apples	91	3	35	NA	61	18	0.07
Citrus	72	2	22	NA	47	24	0.08
All Crops	6	35	17	51	0.9	14	100

[1] NA means not available.

Table 3   Some examples of percentages of crop acres treated, of pesticide amounts used on crops and of acres planted to this crop (USBC 1973; USDA 1975).

These estimated dollar losses increase 3% when all nonfood crops such as cotton, tobacco, hay, and pasture are excluded from the analysis. Based only on food crops, the loss of crops grown without pesticides is estimated to increase from 8% to 11% (Pimentel et al. 1978b).

Crop losses without pesticides were also evaluated based on food and feed energy, expressed as kilocalories (kcal). Losses to insects, diseases, and weeds in crops grown without pesticides but using some alternative controls were estimated to increase only 1% (Pimentel et al. 1978b). When nonfood crops were excluded, the increased loss of crops grown without pesticides was only 4% in food calories.

Based on these estimates of increased food energy losses, 1% for total crops or 4% for food crops, there would be no serious food shortages in the United States if crops were not treated with pesticides. Although the supply of food in the nation would be ample, the quantities of certain fruits and vegetables, such as apples, peaches, plums, onions, tomatoes, peanuts, and certain other crops would be significantly reduced. Because of this, some fruits and vegetables that we are accustomed to eating would have to be replaced with others.

Although our food-energy supply would be little affected by the withdrawal of pesticide use, the dollar loss to the nation would be considerable. This would amount to an estimated $8.7 billion loss, including added costs of employing alternative nonchemical controls that would be used if pesticide use were withdrawn (Pimentel et al., 1978b). Considering that current pesticide treatment costs, material and application, are estimated to be $2.2 billion annually, the return per dollar invested in pesticide control is about $4. This agrees well with previous calculations of between $3 and $5 and adds credibility to our analysis of estimated crop losses if pesticides were withdrawn.

ENVIRONMENTAL AND SOCIAL COSTS OF PESTICIDE USE

In calculating the benefits of pesticides at $4 per dollar invested in control, Pimentel et al., (1978b) did not include a dollar value for the "external costs" of human poisonings and the impact of pesticides on the environment. To evaluate the external costs of pesticide use, the relationship we have with our environment must be understood.

Although everyone knows why food is essential, not everyone is aware of why the environment is equally essential to us. We cannot maintain our high standards of health and achieve a quality life in an environment consisting only of our crop plants and livestock. Most of the estimated 200,000 species of plants and animals in the United States are an integral and functioning part of our ecosystem. Many of these species help renew atmospheric oxygen. Some prevent us from being buried by human and agricultural wastes and others help purify our water. Trees and other vegetation help maintain desirable climate patterns. Some insects are essential in pollinating forage, fruit, and vegetable crops for high yields. No one knows how much the population numbers of these 200,000 species could be reduced or how many species could be eliminated before agricultural production and public health would be threatened.

The impact of pesticides on agriculture, the environment, and public health is significant. There are a calculated 109,000 accidental human pesticide poisonings annually in the United States (Pimentel et al. 1978c). An estimated 6,000 individuals are hospitalized (EPA 1974) and an estimated 200 fatalities associated with the poisonings occur annually (EPA 1974). Apparently none of the poisonings and fatalities were due to eating food crops that were treated properly with pesticides. The people especially prone to pesticide poisoning are pesticide production workers, farm field workers and pesticide applicators. Of the field workers and pesticide applicators in the United States, an estimated 2,826 were hospitalized in 1973 because of pesticide poisoning (Savage et al. 1976).

Because of their widespread use, pesticides are consumed by people. In fact, in one study from 93% to 100% of the people surveyed tested positive for one or more pesticides (EPA 1976). Annual studies conducted by the FDA determine the kinds and amounts of insecticide residues in typical human daily diets. The residues of DDT and its metabolites in foods generally have been low (95% below 0.51 ppm). The incidence of contamination, however, was high, i.e., about 50% of the food samples contained minute but detectable insecticide residues (Duggan and Duggan 1973). Residues of the phosphate and carbamate insecticides are generally less persistent than are the chlorinated insecticides. The FDA data suggest, however, that residues of phosphate and carbamate insecticides are beginning to increase in raw products, and therefore also in the total diet. This increase was to be expected after DDT was banned by the EPA in 1972. In addition, the public will continue to be exposed to residues of DDT and other chlorinated pesticides because these persist in the environment.

At present, overall pesticide residue levels appear to be sufficiently low to present little or no danger to human health in the short term. Samples of fruits and vegetables rarely have insecticide residues that exceed 2 ppm (Duggan and Duggan 1973; FDA 1975). For example, of 1,551 samples of "large fruit" only 10 showed residues of 2.8 to 13.5 ppm. Residues ranging from 2.3 to 84.0 ppm were detected in 97 out of 2,461 leafy and stem vegetable samples. Unfortunately, little is known about the effects long-term, low-level dosages of pesticides may have on public health (HEW 1969). Furthermore, the possible interaction between low-level dosages of pesticides and the numerous drugs and food additives the public consumes has not been completely studied.

The ecological effects of pesticides on nontarget species are varied and complex (Pimentel and Goodman 1974; Edwards 1973). For example, some pesticides have influenced the structure and function of ecosystems, reduced species population numbers in certain regions, or altered the natural habitat under some conditions. Some have changed the normal behavioral patterns in animals, stimulated or suppressed growth in animals and plants, or modified the reproductive capacity of animals. In addition some have altered the nutritional content of foods, increased the susceptibility of certain plants and animals to diseases and predators, or changed the natural evolution of species populations in some regions. Because of this great variation in effect, it is necessary to study the impact of individual pesticides to obtain a fair and balanced picture.

Another interesting aspect of the pesticide problem is the fact that the more than 1 billion pounds of pesticide applied in the United States are used to control only about 2,000 pest species. If these pesticides reached only the target species, pollution would not be a concern. Unfortunately however, only about 1% of the pesticide used ever hits the target pests (PSAC 1965). Often as little as 25% to 50% of the pesticide formulation reaches the crop area, especially when pesticides are applied by aircraft (Hindin et al 1966; Ware et al. 1970; Buroyne and Akesson 1973). Considering that about 65% of all agricultural insecticides are applied by aircraft, the risk both to the environment and to public health is great.

Pesticides are potent biocides and in some cases they may adversely affect the physiology of crop plants. Any change in the physiology of a crop plant can either make the plant more resistant or more susceptible to attack by its parasite and predators. Since crop plants that are not physiologically stressed can more eaily resist parasite and predator attack, any chemical that alters normal physiology is likely to increase the susceptibility of the crop plant. This was demonstrated when calcium arsenate was used on cotton. McGarr (1942) reported that aphids on untreated cotton plants averaged 0.91 per 6 cm^2 of leaf area whereas aphids numbered 6.05 on plants treated with calcium arsenate (about 6.7 kg/ha).

Herbicides have also been found to increase insect pest and pathogen problems associated with corn. For example when corn plots were treated with a regular dosage of 0.55 kg, 2,4-D/ha, aphid numbers on the corn averaged 1,679 whereas on the untreated they averaged only 618 (Oka and Pimentel 1976). Corn borer infestation averaged 28% in the 2,4-D treated corn population compared with only 16% in the untreated corn population.

In laboratory investigations of the impact of 2,4-D on the relative resistance of corn plants to pathogens, exposed corn plants were significantly more susceptible to corn smut disease (*Ustilago maydis*)(Oka and Pimentel 1974) and to southern corn leaf blight (*Helminthosporium maydis*)(Oka and Pimentel 1976).

In summary, the main reasons for serious ecological problems with pesticides are: (1) pesticides are biological poisons (toxicants); (2) large quantities are applied to the ecosystem annually; and (3) application technology is inefficient. Much scientific evidence substantiates the concern that pesticides have caused measurable damage to many species of birds, fishes, and beneficial insects.

We estimate that the external cost of pesticide use is $2 to $3 billion annually in the United States (Pimentel et al 1978c). Included in this external cost estimate are: hospitalization costs for 3,000 human pesticide poisonings; costs of several thousand days of work lost because of pesticide poisonings; and additional medical costs for about 100,000 human pesticide poisonings treated as outpatients. Other environmental and social costs included were: several million dollars in direct honey bee losses; reduced fruit crops and reduced pollination from the destruction of wild bees and honey bees; livestock losses; commercial and sports fish losses; bird and mammal losses; natural enemies of pests destroyed, resulting in outbreaks of other pests; pest problems that result from pesticide effects on the physiology of crop plants; and increased pesticide resistance in pest populations. All of these contribute to the "external costs" of pesticides and must be considered in any cost/benefit analysis.

DISCUSSION

QUESTION: *What is your view on drift problems associated with aerial application?*

PIMENTEL: *There is no better means for distributing pesticides in the environment than using aircraft. Now I am not saying that you should not use aircraft; there are many areas where you have to use aircraft. When cotton is mature or when corn is very high, you have to use aircraft. What I am pleading for is that we have to improve the methods of application so that we reduce the drift problem that does occur with aircraft applications and also discourage the use of aircraft applications when they are not necessary.*

QUESTION: *Isn't a helicopter application much more efficient than aircraft?*

PIMENTEL: *No. The main problem is that when you try to put that material in small droplets, it disperses into the atmosphere and then you have problems. It is a little bit more efficient, but not that much.*

QUESTION: *I realize that we are concentrating on the United States here but I wonder if you have any current figures on China's use of pesticides and their effects on predators and crop losses.*

PIMENTEL: *Yes, China is using large quantities of pesticides and they are having some problems associated with the use of pesticides, as are other nations. In other words, in the United States you hear complaints about all our regulations, but go overseas, into Central America and parts of India, where there are no regulations and see the real problems related to the use of these pesticides. It makes you appreciate our regulations.*

QUESTION: I have always felt that plant resistance is one of the important resistances a plant had. How do you see this strategy being approached?

PIMENTEL: Yes, that is true and they are really going to work on it now. But another point to remember is that the plant breeders came in to deal with insect control when all else failed. There were no insecticides available. Wheat and the Hessian Fly is one example, and, to a degree, corn borer; it was not that effective. This was when our colleagues in plant breeding came to our rescue. This has been true about past efforts in breeding plant resistance and I am glad plant breeders are working in this area now. But if you look at diseases, where they did not have the type of control that we have in entomology, they have been highly successful. I would say 95% of our crops have some degree of resistance to plant pathogens and congratulations to our plant breeders.

REFERENCES CITED

Adams, M.W., A.H. Elingboe, and E.C. Rossman. 1971. Biological uniformity and disease epidemics. *Bio. Sci. 21:1067-1070.*

Beroza, M., L.J. Stevens, B.A. Bierl, F.M. Philips and J.G.R. Tardif. 1973. Pre- and postseason field tests with disparlure, the sex pheromone of the gypsy moth, to prevent mating. *Environ. Entomol. 2:1051-1057.*

Berry, J.H. 1978. Pesticides and energy utilization. Paper presented at AAAS Ann. Mt., Washington, D.C. February 17.

Birch, M.C. (ed.) 1974. Pheromones. North-Holland Publ. Co., Amsterdam. 495 pp.

Bonsma, J.C. 1944. Hereditary heartwater-resistant characters in cattle. *Farming in S. Africa 19:71-96.*

Buroyne, W.E. and N.B. Akesson. 1971. The aircraft as a tool in large-scale vector control programs. *Agr. Aviat. 13:12-23.*

Cartter, J.L. and E.E. Hartwig. 1963. The management of soybeans. pp. 162-226 in The Soybean: Genetics, Breeding, Physiology, Nutrition, Management. A.G. Norman, ed. Academic Press, New York.

Cramer, H.H. 1967. Plant protection and world crop production. *Pflazenschutznachrichten. 20(1):1-524.*

Dahms, R.G. 1948. Effect of different varieties and ages of sorghum on the biology of the chinch bug. *J. Agr. Res. 76(12):271-288.*

Day, P.R. 1973. Genetic variability of crops. *Ann. Rev. Phytopathol. 11:293-312.*

Duggan, R.E. and M.B. Duggan. 1973. Pesticide residues in food. pp. 334-64 in Environmental Pollution by Pesticides. C.A. Edwards, ed. Plenum, London.

Edwards, C.A. (ed.) 1973. Environmental Pollution by Pesticides. Plenum, London.

EPA. 1974. Strategy of the Environmental Protection Agency for controlling the adverse effects of pesticides. Office of Pesticide Programs, Office of Hazardous Materials, Washington, D.C..

EPA. 1976. Human monitoring program. *Pest. Monitor. Quart.*
Rep. #6.

Farkas, S.R., H.H. Shorey and L.K. Gaston. 1975. Sex pheromones
of Lepidoptera. The influence of prolonged exposure to pheromone
on the behavior of males of *Trichoplusia ni. Environ. Entomol.*
4:737-741.

FDA. 1975. Compliance program evaluation. Total diet studies:
FY 1973. Food and Drug Administration. Bureau of Foods, Washington,
D.C.

Frankel, O.H. 1971. Genetic dangers in the green revolution. *World*
Agr. 19:9-13.

HEW. 1969. Report of the Secretary's Commission on Pesticides and their
Relationship to Environmental Health. U.S. Dept. HEW, U.S.
Govt. Print. Off., Washington, D.C.

Hindin, E., D.S. May, and G.H. Dunstan. 1966. Distribution of insecticides sprayed by airplane on an irrigated corn plot. pp.
132-45 in Organic Pesticides in the Environment. *Amer. Chem. Soc.*
Publ.

Lupton, F.G.H. 1977. The plant breeders' contribution to the origin
and solution of pest and disease problems. pp. 71-81 in Origins
of Pest, Parasite, Disease and Weed Problems. J.M. Cherrett and
G.R. Sagar, eds. Blackwell Scientific Publications, Oxford.

McGarr, R.L. 1942. Relation of fertilizers to the development of the
cotton aphid. *J. Econ. Entomol. 35:482-483.*

Metcalf, C.I., W.P. Flint and R.L. Metcalf. 1962. Destructive and
useful insects. McGraw-Hill, New York. 1087 pp.

Moore, W.F. 1970. Origin and spread of southern leaf blight in
1970. *Plant Dis. Reptr. 54:1104-1108.*

Nelson, R.R., J.E. Ayers, H. Cole and D.H. Petersen. 1970. Studies
and observations on the past occurrence and geographical distribution of isolates of Race T of *Helminthosporium maydis. Plant*
Dis. Reptr. 54:1123-1126.

Oka, I.N. and D. Pimentel. 1974. Corn susceptibility to corn leaf
aphids and common corn smut after herbicide treatment. *Environ.*
Entomol. 3(6):911-915.

Oka, I.N. and D. Pimentel. 1976. Herbicide (2,4-D) increases insect
and pathogen pests on corn. *Science 193:239-240.*

Pathak, M.D. 1970. Genetics of plants in pest management. Conf.
Principles Pest Management, March 25-27, Raleigh, North Carolina.

Pimentel, D. 1976. World food crisis: energy and pests. *Bull.*
Ent. Soc. Am. 22:20-26.

Pimentel, D. and N. Goodman. 1974. Environmental impact of pesticides.
pp. 25-52 in Survival in Toxic Environments. M.A.Q. Khan and
J.P. Bederka, Jr., eds. Academic Press, New York.

Pimentel, D., C. Shoemaker, E.L. LaDue, R.B. Rovinsky and N.P. Russell.
1978a. Alternatives for reducing insecticides on cotton and
corn: economic and environmental impact. Report on Grant No.
R802518-02, Office of Research and Development, Environmental
Protection Agency.

Pimentel, D., J. Krummel, D. Gallahan, J. Hough, A. Merrill, I. Schreiner,
P. Vittum, F. Koziol, E. Back, D. Yen and S. Fiance. 1978b.
Benefits and costs of pesticide use in U.S. food production. Manuscript.

Pimentel, D., D. Andow, R. Dyson-Hudson, D. Gallahan, M. Irish, S. Jacobsen, A. Kroop, S. Moss, I. Schreiner, M. Shepard, T. Thompson, and W. Vinzant. 1978c. Environmental and social costs of pesticides. Manuscript.

PSAC. 1965. Restoring the quality of our environment. Report of Environmental Pollution Panel, President's Science Advisory Committee, The White House.

Roane, C.W. 1973. Trends in breeding for disease resistance in crops. *Ann. Rev. Phytopath. 11:463-486.*

Roelofs, W., 1975. Manipulating sex pheromones for insect suppression. *Environ. Letters 8:41-59.*

Roelofs, W.L., E.H. Glass, J. Tette and A. Comeau. 1970. Sex pheromone trapping for red-banded leaf roller control: theoretical and actual. *J. Econ. Entomol. 63;1162-67.*

Savage, E.P., T. Keefe, and G. Johnson. 1976. The pesticide poisoning rate is low. *Agrichem. Age May:15-17.*

Smith, H.H. 1971. Broadening the base of genetic variability in plants. *J. Hered. 62:265-276.*

Stern, V.M. 1969. Interplanting alfalfa in cotton to control lygus bugs and other insect pests. *Proc. Tall Timbers Conf. Ecol. Anim. Contr. Habit. Mgmt. 1:55-60.*

Stevens, N.E. and W.O. Scott. 1950. How long will present spring oat varieties last in the central corn belt? *Agron. J. 42;307-309.*

Sweetman, H.L. 1958. The Principles of Biological Control. William C. Brown Co., Dubuque, Iowa.

Thurston, H.D. 1973. Threatening plant disease. *Ann. Rev. Phytopathol. 11:27-52.*

USDA. 1965. Losses in Agriculture. U.S. Department of Agriculture. Agr. Handbook No. 291, Agr. Res. Serv., U.S. Government Printing Office.

USDA. 1968. Extent of farm pesticide use on crops in 1966. Agr. Econ. Rep. No. 147, Econ. Res. Serv.

USDA. 1970. Quantities of pesticides used by farmers in 1966. Agr. Econ. Rep. No. 179, Econ. Res. Serv.

USDA. 1972. Extent and cost of weed control with herbicides and an evaluation of important weeds, 1968. Econ. Res. Serv.

USDA. 1975. Farmers; use of pesticides in 1971...extent of crop use. Econ. Res. Serv., Agr. Econ. Rep. No. 268.

USBC. 1973. Census of Agriculture, 1969. Vol. 5. Special Reports. Parts 1, 4-6. U.S. Govt. Print. Off., Washington, D.C.

van den Bosch, R. and P.S. Messenger. 1973. Biological control. Intext Educational Publishers, New York.

van der Plank, J.E. 1968. Disease Resistance in Plants. Academic Press, New York.

Walker, J.C., R.H. Larson and A.L. Taylor. 1958. Diseases of cabbage and related plants. USDA, Agr. Handbook No. 144.

Watson, I.A. and N.H. Luig. 1963. The classification of *Puccinia graminis* var. *Eritici* in relation to breeding resistant varieties. *Proc. Linn. Soc. New South Wales 88:235-258.*

Ware, G.W., W.P. Cahill, P.D. Gerhardt, and J.M. Witt. 1970. Pesticide drift. IV. On-target deposits from aerial application of insecticides. *J. Econ. Entomol. 63:1982-1983.*

Whitman, R.J. 1975. Natural control of the European corn borer *Ostrinia nubilalis* (Hubner), in New York. Ph.D. Thesis, Cornell University, Ithaca, New York.

Wolfe, M.S. 1968. Physiological race changes in barley mildew 1964-67. *Plant Pathol. 17:82-87.*

PART TWO
PEST MANAGEMENT IN THE INTERIOR WEST

INTRODUCTION

*I would love to call the local seed dealer or the extension
service and say "bring me out about 10 cases of such and such
insect, I have corn rootworm." I don't like going out there
and mixing up Parathion to kill these insects. It's dangerous
to me, the environment; it's dangerous to everything.*

*I believe we need to have more development on cultural and bio-
logical controls in the region. We just don't have enough of
this type of information to take to the grower and say "this
is a cultural or biological program you can establish in your
field." As I see it, until we get some better information,
we are going to have to rely on pesticides in the immediate
future. We're not blessed with many alternatives at the present
time.*

These two quotes from the Pest Control Strategies Conference by a
Colorado farmer and Extension professor exemplify the state of pest
management in the region. Unlike California which has been very active
in developing alternative pest control strategies since the early 1900's,
integrated pest mangement is still in its infancy in the Interior West.
Many representatives from regional agencies, organizations, farmers and
ranchers at this two-day conference expressed the need to make the transition to more environmentally and economically sound pest management
programs in the region.

As the papers in this proceedings indicate, integrated pest management
is not a panacea that will immediately cure the nation's pest problems.
Rather, it is a methodology that needs to be applied to specific crops,
their pests and the particular environment in which the crop is grown.
It was the aim of the planners of this conference to bring the national
expertise in alternative controls into the region to introduce new approaches for dealing with pest problems. Part Three explains some
alternative control strategies that have been shown to be viable in
various parts of the country and Part Four discusses the implementation
of IPM programs.

Using excerpts from panel discussions, presentations and papers, Part
Two focuses on some of the pest problems and current control strategies
in the Interior West. We have tried to provide a cross-section of views
on pest management from diverse perspectives. This synopsis of pest
management in the Interior West is important in assessing the task of
implementing more comprehensive IPM programs in the region. Several programs will be discussed that are at various stages of development; for
example, a biological control program in orchards on the west slope of
Colorado has been ongoing since 1946. We hope this approach will be of
benefit not only to the region but to those who wish to see IPM implemented in other regions of the country.

REGIONAL PEST PROBLEMS

Colorado ranks first in the country in sheep and lambs on feed, third
in sugar beet production, fifth in cattle on feed and pear production,
seventh in sorghum and corn for silage, thirteenth in winter wheat,
and fifteenth in alfalfa hay production. Winter wheat and corn are the
two largest crops in terms of dollar value and total production (1977
Colorado Agricultural Statistics, Colorado Department of Agriculture).

Crop Pests

*Excerpts from transcription of conference working session: William
Hantabarger, session leader.*

Approximately 80% of the corn in Colorado is grown continuously on the
same ground. Where corn is grown continuously, corn rootworm is usually
a problem. In Colorado, corn accounts for the majority of insecticide
applications at the present time. We recommend a planting time or
post emergence soil insecticide as an insurance treatement against corn
rootworm because we have no accurate way of telling if a particular
field needs control for the coming year. We have begun to look at an
adult control program as an alternative to treating with soil insecticides each year. This will involve close field monitoring for adults
and selective spraying.

QUESTION: Doesn't spraying for adults lead to secondary pest resurgences?

*ANSWER: Yes, Banks grass mite is our second most important pest in
corn and we know we are affecting the mite's parasites and predators and
thus creating our own mite problems with an adult spray program. This
is also true when controlling western bean cutworm which is rapidly
spreading in our region.*

QUESTION: Have you tried Bacillus thuringiensis on the cutworm?

*ANSWER: Yes. We have conducted field trials along with other pesticides
but it doesn't look too promising yet.*

*COMMENT: There are difficulties with field trials for Bacillus according
to Texas A&M studies. To get a fair test of MPV or Bacillus you have
to use large blocks in areas of low chemical use. A combination of beneficial insects and microbials gives an additive effect that you do not
see in small plots because of pesticide drift which biases against the
microbials.*

The major pest problem on sorghum is the sorghum greenbug. This is an aphid that has a very toxic feeding secretion. We base our economic threshold on plant size. Resistant hybrids of sorghum are being researched.

There is one pest problem on small grains which is rather detrimental as far as the environment is concerned, the pale western cutworm. Populations increase in dryland wheat following a series of dry years on the high plains. The only insecticide presently registered is Endrin which is quite toxic to non-target organisms, particularly birds. We desperately need some alternatives for this pest.

As an extension entomologist in Colorado, I feel we need better information on economic thresholds. More information is needed on the basic biologies and life cycles of major pest species. We also need more development and more support for cultural and biological control investigations.

Noxious Weeds

Excerpts from transcription of conference working session: Eugene Heikes, session leader.

The two most important weed problems in Colorado are Canadian thistle and field bindweed. Of less importance in our rangeland and cropland are: bull thistle, Scott's thistle, musk thistle, Russian knapweed, and the poverty weeds.

We do not have adequate controls for our perennial weeds. We can control most weeds in Colorado but the economics are prohibitive in many cases. We do not have good economical controls.

We have made several surveys in Colorado and we feel that weeds cost our farmers about $2000 per farm. Weed control of Canadian thistle and field bindweed costs farmers in the neighborhood of $100 an acre.

Many of our perennial weed infestations have started from contaminated seed or feed. It was just a few years ago that the Colorado Department of Agriculture placed embargoes on contaminated seed entering the state.

One of the control methods for Canadian thistle which we strongly recommend is a cultivation program involving frequent tilling. We also recommend the use of good competitive crops. Some of the herbicides recommend are 2,4-D, and 2,4-D and Banvel combinations. Basically, the herbicides simply starve out the root system by destroying the top growth. This may take several years. When you consider that the seeds of some weeds are viable for approximately 30 years in the soil, it is clear that weed control is a long-term project.

The tremendous root system of Canadian thistle is one reason that I feel makes biological and cultural control difficult. There is enough storage in the roots of Canadian thistle to last three years without much top growth. If control insects die off or migrate within three years then the weed will just grow back.

There is probably more field bindweed than Canadian thistle in crop-lands. Field bindweed causes the biggest problem in dryland wheat growing areas. Yields have been reduced in some cases by as much as 50%.

Weed control should be considered in any farming plan. I see chemical weed control as a supplement to good farming practises.

QUESTION: Do you have the enabling type of legislation in Colorado to develop weed control districts?

ANSWER: Yes, we have tried to get a state-wide law but with no success. We do have an enabling type legislation where counties as a whole and districts within a county can form weed districts. We do have a few of these districts within the state. They would be better replaced by a state-wide system.

QUESTION: Do you have any recommendations on a system that would operate on a no or low-tillage program in wheat farming?

ANSWER: Minimum tilling is good but it encourages weeds because the soil is not turned over which permits the weeds to survive. In programs of low-tillage, weed problems have been severe.

QUESTION: Have studies been run in relation to chemical weed tillage also?

ANSWER: Not as much as could be. Most annual weeds associated with minimum tilling can be destroyed by use of 2,4-D. We will probably have to use more herbicides with minimum tillage than we have in the past.

Range Insect Pests

Excerpts from transcription of conference working session: Lowell McEwen, session leader.

For many years Aldrin and Dieldrin were the main pesticides used for grasshopper control in range. They were, of course, very toxic to wildlife, particularly birds. Malathion which is presently used, is quite effective for grasshopper control and has little effect on wildlife. Malathion is the only chemical that the USDA recommends for large spray operations.

Grasshopper spray operations probably involve several million acres of rangeland a year. The typical spray program involves an entire block of land ranging from 100 to 500 thousand acres.

During the last ten years, the USDA Range Insect Control research team in Boseman, Montana has been developing a biological control program for grasshopper control. The control agent is an endemic protozoan, *Neosema locustae*. *Neosema* feed on the grasshopper fat bodies and it

24

takes about two weeks from the time the grasshopper has been-infected
to kill the grasshopper.

The spores which are extracted from laboratory bred populations are
broadcast as an aerial spray. Each grasshopper contains enough spores
to treat two acres. The cost is about 25% to 30% of the cost for chemical
application.

One objection to the program is that there is not a quick knock-down
of the grasshopper population. This biological control program will
have to be fit into a total pest management system in which potential
infestations can be treated with *Neosema* before the infestations become
too severe. The USDA research team is also looking into the possibility
of a combination of a low dosage Malathion treatment for immediate knock-
down with the *Neosema* spore release.

Urban Pests

*Excerpts from transcription of conference working session: Byron
Reid, session leader.*

The word sanitation seems to come up in almost every pest control area.
In urban areas it is of vital importance. In the ornamental area and
the fruit growing areas, sanitation has also been stressed.

A major point that was stressed during our session on urban pest control
was the overuse or misuse of pesticides by homeowners who buy the chemicals
off the shelf and use them without having proper knowledge of their use.
I see consumer education as an important facet of developing sound pest
mangement programs in urban areas.

In the urban and ornamental areas we stressed the need to use pesticides
on an as-needed basis. Everyone in the room felt that we were looking
forward to the time when better alternatives are available.

Forest Pests

*Excerpts from transcription of conference working session: Robert
Stevens, session leader.*

The mountain pine bark beetle is the major forest pest in the Rockies.
The mountain bark beetles lay their eggs just below the bark layer of
Ponderosa pines where the larvae form galleries to feed. The pine bark
beetle acts as a carrier for a blue-stain fungus which destroys the
transporting tissues of the tree and causes the tree's eventual death.

Forestry is not a big industry in this part of the country so there
isn't alot of industry interest in the forest pest problems. The more
important values of forests in this region are recreation, aesthetics,
watershed and wildlife management.

We have to realize the lack of water on the Front Range. Though the area had a history of heavy lumbering around the turn of the century the rate of turnover for the forest is too slow for viable lumbering operations.

Forestry has some similarities with agriculture. We have monocultures; contiguous stands of the same species. Many areas, however, have mixed stands with diverse age-structure and, as you might expect, fewer pest problems. In forestry, of course, we don't plant the crop in April and harvest the crop in September. In fact, in most cases we haven't planted it at all and have no plans for harvesting.

One difficult aspect of forestry pest management is the time it takes to grow a crop and how that affects certain techniques like breeding resistant hosts. In the case of the mountain pine bark beetle, it takes approximately 50 years before the pines are even susceptible to attack. You can screen a tree for a particular turpine content which seems to be resistant to the pine bark beetle and not know for 50 years if you made the proper selection. Also, because the insect populations turn over at least once a year, the chances are clearly in favor of the insect changing faster than the host tree species.

COMMENT: Isn't the longevity and stability of the forest ecosystem a factor that is also in your favor. For example, the use of natural enemies has been shown to be more successful in stable systems such as orchards than in croplands where radical changes occur each season.

The result of that early logging along the Front Range has been the development of a monoculture of Ponderosa pine, even-aged and very dense. This is the optimal situation for infestations of mountain pine bark beetle.

In dealing with forest pests we like the approach of fiddling with the hosts, a process known as environmental and habitat management, rather than trying to control the pest population.

Our program for fighting the pine bark beetle is a rather unsophisticated IPM program. It involves the use of pesticides for direct control coupled with the cultural approach of thinning stands. Ideally, this type of cultural approach would be best done in advance of the problem. Unfortunately, we have, by necessity, to attempt control right in the midst of the problem.

QUESTION: Is there anything being done about the possible use of parasites for a control program?

ANSWER: The parasites are of little importance; the predators of slightly more importance. Clerid beetles and woodpeckers are a factor but they are not dependable.

COMMENT: Some studies have indicated that there are fairly sizable populations of parasites in moist environments in Colorado, along ravines and canyons.● On the exposed knolls or in xeric environments, the populations were much lower. So it appears that the parasites can be effective but in very selective situations.

26

COMMENT: _One of the problems in the use of parasites for control programs_
is that the parasites have to eat. _I feel that some releases have failed_
because a basic food source for the adult was not available. _There are_
lots of ichneumons that can find a batch of eggs or larvae beneath the bark
but they must have moist, food sufficient, areas in which to live. _I think_
that the bark beetles are probably much more uniformly distributed than
damage lets us know. _Where there is food and water for the adults, you_
have a healthy population of parasites keeping the infestation down at the
sub-economic level.

QUESTION: _Isn't it true that female pine bark beetles are attracted to_
pheromone traps and if so can't you use that as a control method?

ANSWER: _There has been alot of pheromone work done on bark beetles and_
the results are rather spotty in general. _We haven't been as successful_
with bark beetle pheromones as with the lepidopteran pheromones which are
largely sex pheromones; bark beetle pheromones are aggregating pheromones
which is quite different. _Pheromones are used to sample population densities_
of pine bark beetles.

It will be interesting to see what happens in Rocky Mountain National Park
where there is no control for pine bark beetle. I think they are doing
the right thing up there and the State Forest Service is doing the right
thing by trying to control the beetle near the cities.

VIEWS ON PEST MANAGEMENT

Glen Murray *is a farmer from Brighton, Colorado. He produces grain, corn and alfalfa on an irrigated farm which is used to support the cattle industry in the general vicinity. He is a member of the Rocky Mountain Farmer's Union.*

First, let me begin by giving a little personal background. I have been in this buisness of agriculture directly or indirectly all my life. I was born and raised on an irrigated farm and have received two degrees from Fort Collins in agronomy.

You have all heard about the problems that agriculture is in today. I would like to give a little history of what is actually going on. Today in the United States approximately 4% of the population is involved in agricultural production. That percentage is smaller than anywhere else in the world. The income of the average farmer for this past year was approximately $8,000. By way of comparison, the average factory worker earns approximately $12,000. Now that doesn't sound too drastic in itself, except when you consider that income involves the farmer's entire family. You can see that agriculture today is having some real problems. Another fact is that in farming today, the average farm is valued at a quarter of a million dollars. The average return on that investment is four percent. You can see that unless, as the old saying goes, you marry it or inherit it, you do not get into agriculture today, at least not in the production end.

One point that was brought up today deals with the economics of size. In other words, the big corporation has it made. The point that has not been brought up is that when you look at the efficiency of corporate agriculture, studies indicate it has been much less efficient than the family farm or small farm operations. I question the concept that bigness is better; the statistics don't prove it out.

People, today, are always talking about our standard of living, about how high priced everything is. One should stop and look around the world and make some comparisons. We enjoy a higher standard of living in this country than anywhere in the world. Why do we enjoy this high standard of living? The major reason is the cost of food. We pay less than 17% of our disposable income for food in this country. That is lower than anywhere in the world.

The issue of monoculture versus diversified farming came up this morning. Someone asked "why doesn't everybody grow a little of everything." There are several reasons for this. The main reason boils down to economics again. The producer cannot afford to do that for two reasons.

One is the cost of machinery. Today, a small combine is going to cost you about $60,000. How many ten or fifteen acre plots do you have to have to pay for that $60,000 machine ? You can say, "Let's use labor and then you don't have to use that machine," but then where do you get the labor and how do you pay the $5 or $6 per hour that it takes to compete with the urban market ? You just can't do it.

This conference, however, is concerned with pest management and I would like to explain my viewpoints. Mr. Tweedy defined integrated pest management very well and I agree with him. I hate to use chemicals just as much as anyone else but unfortunately, at the stage of the game today, it is one of those necessary evils. I am on an irrigated farm where highly intensive agriculture is conducted and for me that is the way I see it.

The people that profess not using any chemicals in agriculture don't look far enough down the road. I would love to call the local seed dealer or the extension service and say "bring me out about ten cases of such and such insect, I have corn rootworm." I don't like going out there and mixing up Parathion to kill these insects. It is dangerous to me, the environment; it is dangerous to everything. The worst part of it is that it does not always work. I think, presently, it is basically a re-search question and this is where I am going to lean on the research people. There needs to be alot of work done. I think there is some real potential in cultural control programs, biological control programs and in resistant varieties.

The one thing I would like to leave with you is that today in the United States, the American farmer feeds this country and a good portion of the world. Now, if we want to take a step backwards and eliminate chemicals from agriculture, that's fine. But I want you to think about the impli-cations of this. Food is really the only thing that this country can export to balance our trade deficit. If we eliminate chemicals completely, we are going to have problems.

Thomas Lasater *is a rancher from Matheson, Colorado. He has developed and expanded the basic herd of one of the two modern breeds of cattle, the Beefmaster.*

Many years ago an old Texas cattle friend was travelling through some of the back country in Mexico and saw this roadside establishment and he thought he would stop in for a cup of coffee. He stepped inside, ordered a cup of coffee and looked off to his right where an elderly Mexican gentleman was having lunch. His plate was completely covered with flies. This old fella was paying no attention to the flies, just happily eating away. Finally the Texas turned to this old Mexican gentleman and said, "pardon me, but don't those flies bother you?" The gentleman replied, "oh no, they eat so little."

Before launching into the main subject today, I would like to say a word about land ownership. We do not and I am sure many of you do not look upon the ownership of land in the same category of owning a home or owning an automobile. We are merely temporary custodians of the land. It is up to each one of us to pass on to the next succeeding generation a better piece of land in better condition than we received when we took over.

The same philosophy should also apply to the environment. It is, of course, of paramount importance that each of us should familiarize ourselves with our own environment. Secondly, we should take definite steps in our daily lives to do something about it, not just talk about it.

At our ranch in Matheson, Colorado, we have never used herbicides and in 1949 discontinued the use of all pesticides. We used to spray in December and February for lice. An interesting point to us is that we have fewer lice problems on our cattle in the wintertime today than we did when we sprayed. As we all know, each animal, including insects, has its own natural enemies and apparently we were killing more of the louse's natural enemies than we were lice. Once we quit spraying the natural enemies took over and have cut down the lice population substantially.

Another interesting thing is the matter of predators. When predators and their victims are left to their own devices, they will seek a balance. For instance, when we first came to Colorado, we literally had thousands of thousands of rabbits on the ranch. We had three different species of rabbits. We moved in with a no-hunting, no-trapping, no-poisoning policy. Nature immediately brought in hordes of coyotes and they cut the rabbit population down to a normal size. As soon as they finished the job the coyotes left, leaving a stand-by crew to maintain the situation.

We were unfamiliar with prairie dogs when we arrived and the neighbors and the county agent told us we should kill the prairie dogs. We had one prairie dog town and, unfortunately, we took their advice and poisoned them. Several years later, I was riding across the pasture with a friend and he pointed to where this prairie dog town had been and asked, "why is the best grass in this pasture right over there?" I replied, "that's where the prairie dogs were." Since then we have imported two different batches of prairie dogs and they have refused to settle on the Lasater ranch. They all go off to the neighbors and I am sure they get poisoned.

A wealthy Texas oilman bought an island off the Texas coast so he and his friends could go quail hunting, a regular quail preserve. There were alot of hawks out there so he sent these hunters to obliterate the hawks. The quails vanished. They found out that the hawks were eating the mice and thus keeping the mice from eating the quail eggs. When there were no hawks, the mice multiplied, ate all the quail eggs and that ended the quail hunting. These examples show that it is best to leave nature alone.

We often talk about the balance of nature. Actually, nature operates in imbalances in the short-time frame. Some examples are tornadoes, cyclones, floods, and droughts. In the long-time frame it all balances out. A perfect example of this was the migratory practices of the buffalo in the early days. They moved from the panhandle of Texas up into eastern Colorado in these huge herds of four or five thousand and they would absolutely decimate the country as they migrated. They ate grass, shrubs, trees, everything; what they did not eat they would trample down. The following year the migratory route of the buffalo would be further to the east or west and they wouldn't come back to the same spot for at least two or three years, giving nature a chance to restore the land.

Pauline Plaza is on the staff of the Western Environmental Science Program of the National Audubon Society. Ms. Plaza received her M.S. in Wildlife Biology from Cornell University.

The National Audubon Society is involved with a large variety of issues across the country. At our office here in Lakewood, Colorado, we cover mostly the western issues: everything from oil shale to golden eagles and things in between. Some of the comments I am about to make may seem strange to you if you are thinking only in terms of crop pests, but if you think in terms of prairie dogs versus cattle, or golden eagle depredations on lambs, they make more sense.

I would like first to describe a couple of the major environmental goals of the Society. Most people know roughly what they are but seldom see them in print. The two of interest to us here are: the conservation of wildlife and the natural environment; and the prevention and abatement of environmental pollution in all its forms. These are very general goals that the Board of Directors formulated some years ago. Included under the second of these goals is "advocating biological and integrated pest control measures." This is our attitude towards Integrated Pest Management. However, the Audubon directive adds, "while working to eliminate persistent and highly mobile pesticides and toxic substances that poison the food chains of natural ecosystems." So, on the one hand we are very interested in <u>biological</u> <u>control</u>. On the other hand, we discourage the heavy and exclusive use of chemical pesticides as practiced in American agriculture in the past. I think agriculture is in a period of transition right now.

The Audubon Society has a long record of opposition to the exclusive use of chemical pesticides. The reason for our position is not hard to understand. We are concerned with environmental and social costs that are not usually included in a cost/benefit analysis of pesticide use. These are hidden costs that go back twenty or thirty years. They were certainly not foreseen then, but we do know about them now.

Environmental costs include the massive disruption of complex biological communities and the loss of function of some of these communities. For example, early applications of DDT essentially eliminated bird and insect

31

predators in certain areas and threw the whole local food web out of
kilter. The end result, as we all know, was an increase in the number
of pest individuals because their natural predators had been destroyed.
What the other ramifications of removing species from their communities
are, we are just beginning to find out.

Consider also that intact natural communities perform valuable services
for man -- e.g. flood and erosion control, air and water filtration,
and moderation of temperature extremes, among others. You can see that
these communities are quite important to human welfare, whether we
realize it or not. When we disable them we must replace their functions
with man-made systems at enormous cost. Apart from their functional
importance, natural communities have aesthetic and recreational values
which we are just starting to recognize and, in some cases, quantify.

Other environmental costs that the Society is concerned with are the
direct losses of fish and wildlife and, particularly, the indirect losses
through the concentration of persistent, toxic chemicals along the food
chain. The decline of the peregrine falcon is a good example. Most of
you have heard this story before, but I think it's worth repeating. The
peregrine was a fairly widespread breeding bird up to 40 years ago. It
has been eliminated as a breeding species in the eastern United States
and severely decimated in the West due primarily to the use of DDT. It
survives in the east only because of captive breeding and reintroduction
efforts. In Colorado only six out of 30 historical eyries are still
active, an 80% reduction. The problem is not destruction of habitat -
the habitat is mostly intact - but rather eggshell thinning and consequent
low reproductive success. DDT has been incriminated as the chief
agent of this process.

So - how do you put a cost on a thousand peregrine falcons and the disruption of the natural community in which the peregrine is on the top
of the food chain? These kinds of costs are hard to quantify, though
you could, for instance, find out how much is being spent each year to
breed and restore peregrines to the wild. One estimate is $2000 per
young bird.

The Audubon Society is also concerned with the social costs of pesticide
use. There are first the direct costs of human life due to mistakes in
application of pesticides. There is also the deterioration in human
health and efficiency due to sublethal doses of pesticides. Then there
is the direct cost to the farmer of increased pesticide use. As the
natural enemies of pests are eliminated by extensive spraying programs
and the few naturally immune pest individuals begin to breed, secondary
pest outbreaks occur, worse than the first. As this spiral accelerates,
yields start to drop, as documented cases prove. The end result is an
increase in direct costs to the farmer, with the possibility that he will
be driven out of business by the combination of increased costs of pesticides and reduced crop yields.

There is something else we have not quantified yet. That is the long-term cost to society, in health care for example. Or the eventual costs of an agricultural system that relies on chemical pesticides and destroys the natural communities that could provide a buffer zone for predator species. We just do not know what those costs are.

I think you can see that the Society is very supportive of integrated pest management. Some aspects appeal to us more than others - biological controls, cultural diversification, less use of marginal land for farming and livestock grazing and more emphasis on lands that are well adapted for these purposes.

When a pest problem arises, we keep several considerations in mind:
1. Is there really a problem? This may seem a strange question if you are standing out in the middle of your field and insects are crawling around you, but in other cases this is a valid question. For example, are vertebrate predators on livestock really a problem? Were the huge blackbird roosts in Tennessee and Kentucky a real hazard to human health? The problem is usually the icing on the cake; all you see is the icing and you may not see the basic problems underneath.
2. We always consider the safety to humans of the control method, whether it is a pesticide or not.
3. Is the control method specific? Is it aimed just at the pest? We do not support the use of broad spectrum control methods because they play too much havoc with natural communities.
4. Is the method effective and efficient? Like everyone else, we don't like to see taxpayers' money spent on methods that don't work or which can only lead to escalating costs. We protest some measures on that basis.
5. Most importantly to us, and I think to a lot of environmental organizations, is the question of what the long-term impacts of pest control methods will be. Again this stems from our concern for the maintenance of natural communities or the biosphere, if you prefer to call it that. I cited the loss of the eastern peregrine falcon population as an example. This is a long-term impact that we did not foresee but which we now know to have been directly related to the use of DDT and other chlorinated hydrocarbons. We don't know about other methods, but at least we know there can be this kind of effect.
6. Then lastly, a question we are concerned with is: what are the synergistic effects of the control action? Sometimes a single action can have multiple effects. Draining a swamp to control mosquitoes has impacts on flood control, ground water levels, water quality, wildlife abundance and a number of other factors. When a wetland is drained, you are not just getting rid of mosquitoes but rather affecting many other parts of the environment. The question arises whether the benefit, in this case the projected absence of mosquitoes, is worth the costs which you may not be able to quantify or even predict. These multiple effects must be kept in mind.

Another of the Society's major concerns is the exportation of American agricultural techniques to foreign countries. Audubon does quite a bit of work in Mexico. We have done biological surveys in Central and South America in which we were concerned mainly with migratory birds such as waterfowl, shorebirds, songbirds and some raptors. The United States is still exporting hard pesticides to these countries. Their use is having disastrous effects not only on the wildlife and natural ecosystems of these countries but also on the economic structure and social stability. We really question whether we should export our technology abroad. You may say it increases food yields. This may be true temporarily, but in the long run the effectiveness of chemical control has been shown to decrease. Are we morally justified in exporting a process which we now know will not work in the long run?

One thing we would like to see is exportation of Integrated Pest Management techniques. If we are going to export technology, let's at least export an effective and efficient one.

The Society's actions include legislative lobbying by our office in Washington. We also have direct contacts with the executive branch and the agencies having jurisdiction over pest control. We also have quite a large public education campaign. Most of you have probably seen our magazine.

To summarize, our basic concern is the long-term effects of certain pest control methods on ecosystems, on wildlife and on humans. All life ultimately depends on the functioning of complex biological communities. Audubon's concern is the maintenance of the whole complex of interdependent species and the abiotic environment, not just crop ecology. We realize that choices sometimes have to be made between human health and a temporary disruption of natural communities, but such cases are relatively rare.

We urge that biologically sensible methods be used to control pests; less damaging techniques than the blanket application of broad-spectrum, toxic persistent chemicals. I think that American agriculture will eventually move away from this. The sooner the better as far as we are concerned.

Wayne Bain *is the executive secretary for the Mesa County Peach Administration Committee which is a grower organization for fruit growers on the west slope of Colorado.*

I would like to make a few recommendations regarding pest management. I made a telephone call yesterday and I noticed an interesting thing that I think applies to the problem that we may have with growers accepting IPM. When I was making the call, I noticed they had a small blackboard beside the phone to prevent people from writing on the wall. Someone was intelligent enough to provide a good alternative to writing on the wall. This is the first thing that I believe that growers need: a low-cost efficient alternative method of control to pesticides.

The second recommendation is to set up a procedure for faster and better chemical evaluations by those concerned. I had not realized, until Mr. Reese explained yesterday, that court cases were one reason for the delay in chemical registrations. Another recommendation would be a more logical review of the risk/benefit ratio in considering the use of restricted chemicals.

I read recently that it had been suggested to form a group called the Inter-agency Regulatory Liaison Group with representatives from OSHA, EPA, FDA, and the Consumer Product Safety Commission. What bothers me about this, and Mr. Reese also commented on this, is that it is very difficult to have multiple agencies working on a common problem. It is also difficult to have one involved with regulation and another charged with the responsibility of compliance. I think all this responsibility should be centered in one single body. Mr. Reese stated that the regulatory aspects of pest management had been pulled together more or less under EPA but he made the point that the aviation group still monitors one segment.

The last recommendation that I would like to make is to use some diplomacy in tackling grower relationships. A good government/grower relationship would result in a more willing compliance as opposed to the use of force. Growers are getting highly sensitive about the number of regulations that they have to operate under.

I'll close with a quote, I do not know the author. It states, "private initiative often works hardest when government intervenes least but seldom have we given it a real chance to work. It is a rare case when the passing of a law cures the problem."

Allan Jones *is a fruit grower from western Colorado producing peaches, apples and pears.*

I represent some 200 peach producers in a marketing association in western Colorado. This grower organization is highly organized and can do many things that other groups cannot do because each producer within this group has to pay his fair share. Voluntary methods have never done the job.

Pest management, in my opinion, is a huge ball of wax and there is no accurate way to evaluate what is good and what is harmful to mankind. I see pest control as a war, a battle to grow enough food and fiber so that people can live well. I have used chemicals for some thirty years. As a fruit grower, I have ridden a spray rig for many years, and although I have a lot of gray hair now, I am still around. Every one of my children raised on our orchards is healthy. Lack of caution and carelessness can be a problem in the application of chemicals but I doubt that there is very much proof that a chemical properly used has killed anyone. I am sure that this is debatable and I will be taken to task for what I say.

As a producer, and I know a good many producers across the country who feel the same way, I am somewhat reluctant to believe our government any more. I dislike saying this but that is how producers see it. We get statistics till they are running out of our ears and often one set of statistics disproves the other. Agencies fight among themselves, one agency not knowing what the other is doing. We don't know what they are doing. I hope that as time goes on we will have less government in our business.

To get to the point of today's discussion, we have been involved with biological control for some 32 years in the industry. It is not anything new to us. Around 1946, the peach industry found itself in trouble with the oriental fruit moth. We started to develop a control program with the Colorado Department of Agriculture. We furnished a large part of the money in the beginning. Since that time we have had an integrated program on the oriental fruit moth. We raise a lot of the parasite *Macrocentrus*. The cost of the control is $7 per acre versus $30 per acre for chemicals. You know that we are in business and that we are going to use the best method to control this fruit moth. We use both the parasite and chemicals to get the job done. In our area we are always looking for new control methods. At the present time, the Insectary in Palisade which is the only one in Colorado has five programs in biological control. We are very appreciative of the Colorado Department of Agriculture and its insectary on the west slope for the job they are doing.

We also have another group of people in the state that we are quite proud of and that is Colorado State University. They give us help by writing a booklet each year that we call the "Fruit Grower's Bible." This booklet outlines the different control programs that we might use, the sprays, how much to put on and when to spray. So we have two different organizations that are working with us and each doing a good job.

If you want to keep pests at a minimum, first you need proper sanitation. In our area we have pest districts. The growers monitor themselves and there are also inspectors in each area. If a grower doesn't spray at the right time or lets a pest get out of hand, he is told to take care of it. If he doesn't, somebody else will take care of it for him.

We have used biological controls, predator mites and parasites, for the elimination of oriental fruit moth and we are still using chemicals. There is a gentleman in the audience, Les Ekland, an IPM consultant, who works for me. I pay him a pretty good salary out of my pocket but I think he is probably going to save me money. We use as little chemicals as possible; that is why Les has a job. We used to apply five or six sprays a season for apples. Now, Les watches his pheromone traps and tells us when to spray. Unfortunately, you can get a bad case of ulcers sitting around waiting. We used to spray whenever the University said so, whether we needed to or not. Les has taken all this gamble out of this.

To summarize, we are going to have to work together to get the job done. We don't think that government intervention in our business is the answer to this thing. It is up to us. The program I have listened to these last two days has enlightened me greatly about what our chances are to do a job and use less chemicals doing it.

REGIONAL PROGRAMS
IN INTEGRATED PEST MANAGEMENT

BIOLOGICAL CONTROL PROGRAMS

Albert Merlino

Insectary Section,
Colorado Department of Agriculture

The Colorado Department of Agriculture, long before the term "Pest Management" came into usage, was committed to biological control with the introduction of the Oriental fruit moth, *Grapholitha molesta* (Busck) in 1944. Since the biology of this pest did not respond to chemical controls, a mass rearing program of the parasite *Macrocentrus ancylivorus* was initiated in 1946.

The outgrowth of this initial venture into the biological control field resulted in the Colorado Department of Agriculture, Division of Plant Industry establishing the Insectary Section which is based in Palisade, Colorado. Since that time, working with the United States Department of Agriculture and related agencies across the nation, Colorado has become involved in many beneficial programs which have had significant impact against some important pests in the region.

This has proven to be true with the releases of Oriental fruit moth parasites where there is full cooperation with the peach growers. The parasites are timely released by using the data from trapping the moths by the pheromone method, maintaining records of applied pesticides and calculating the degradation time of toxic residue by the bio-assay method. This program is unique to the peach growing district of Western Colorado and has reduced chemical application requirements from 3 to 1 and in many cases none.

There are 3 to 4 generations of Oriental fruit moth in Colorado. The last generation occurs preharvest when larval damage would be expected to be the greatest. Timely utilization of *Macrocentrus ancylivorus* has eliminated the need for chemical control as evidenced from harvest samples showing 0 to 2 percent damage, averaging .2% for orchards monitored in 1977. The effectiveness of this parasite has been aptly demonstrated. Cooperation, record keeping, analyzing trap data and communication between the peach grower and the insectary is the primary consideration for a successful program.

Monitoring of two-spotted mite, *Tetranychus urticae*, populations and the utilization of the western mite predator, *Typhloromus occidentalis* (Guthion resistant) in fruit orchards has been one of great interest. Results, thus far, indicate that this predator can effectively contol mite populations under properly managed conditions. However, the question remains, does the average grower have the knowledge and tools to effectively evaluate the problem?

In working with the Dutch elm disease problem, *Ceratocystis ulmi* and its vector *Scolytus multistriatus*, a parasitic wasp *Dendrosoter protuberans* imported from France by the USDA and supplied to Colorado for rearing and release, has been successfully established and monitored since 1974. Eventually all counties in the state having populations of this beetle will recieve parasites.

The parasite has effectively overwintered at primary release sites and recovery studies have shown a maximum of 30% parasitism. Studies also indicate that *D. protuberans* will also parasitize *Scolytus rugulosus*, shot hole borer of peach and other fruit trees, as well as the olive bark beetle, *Leperisinus californicus* taken from infested blue ash trees.

Colorado has, over the years, made successful introduction and colonization of other exotic species. Probably one of the first was *Chrysolina quadrigemina* introduced to suppress Klamath weed, *Hypericum perforatum*, in 1956. Klamath weed is toxic to livestock. The weed, especially troublesome in the Rocky flats area of Boulder county, is now well under control by this imported Australian beetle.

More currently, introductions of *Rhinocylus conicus*, a seed weevil from Europe, have been made in an effort to suppress *Carduus nutans*, the musk thistle. Surveys confirm the thistle to be readily established and competing with desirable plant species in 18 counties of the state. Trial releases began in 1974 and establishments have been made in Larimer, Mesa, and Eagle counties. Two sites in Larimer county which were at explosive population levels now show marked reduction of thistle. Collections for sub-colonization to other infested areas of the state were made in 1978 from one Larimer county site. The weevil is expected to be a successful adjunct in suppressing the thistle and relieving costly labor and control costs.

In the forage crop area, the Indian wasp, *Aphidius smithi*, was introduced to assist in suppressing *Illinois pisi*, the pea aphid, which was causing tremendous damage to legume crops. The introduction was successful and collections from established sites were made to other problem areas of the state. Since its introduction, pea aphid complaints have been minimal.

Two new programs now in progress are the rearing of *Liotryphon sp.* obtained from the University of California, which parasitizes *Laspryesia pomonella*, the codling moth, and investigations are underway to determine the suitability of a biological control program utilizing parasites of *Hemileuca oliviae*, the range caterpillar. 1977-78 overwintering trials of *Liotryphon sp.* were successful. This large, docile parasite-predator from Afghanistan may well fit into areas where organic farming is practiced and in areas where pest and ornamental host trees harbor populations of codling moth larva.

Cooperative efforts with Colorado State University, Zoology-Entomology

Department, featured studies of alfalfa weevil parasite, hyper-parasite spectrum, relationship to weevil control and parasite protection with respect to hay height, olfactory response studies to certain sex phero-mones as to whether parasites react favorably to the pheromone of its host and recovery studies of the elm bark beetle parasite, *Dendrosoter protuberans*.

In summary, the Insectary Section of the Colorado Department of Agricul-ture's Division of Plant Industry is committed by law to rear, release, introduce and colonize exotic beneficial insects as they become available through USDA agencies for integrated control of entomophagus and phyto-phagus pest species. Reciprocity in exchanging biological agents is a unique feature enjoyed with other states and countries. Furnishing para-sites of Oriental fruit moth to Russia and Australia highlights the respect Colorado enjoys in the biological control field.

INTEGRATED PEST MANAGEMENT IN ALFALFA

Donald W. Davis

Department of Biology,
Utah State University

Alfalfa was one of the first crops considered for an integrated pest management program. Many of the earlier attempts at integrated control and supervised control were centered on alfalfa and when the major crop systems were selected for detailed pest management and agroecosystem studies about 8 years ago, alfalfa was one of the 6 selected. Alfalfa has features that make it an ideal crop for integrated pest management as well as features that are undesirable.

Favoring IPM, alfalfa has the following:
1. It has a liberal economic injury level. In other words, a significant amount of pest activity can be tolerated.
2. Most plantings are fairly large.
3. In most areas there are relatively few key pests; in the northern Great Basin there is only one on forage alfalfa.
4. It is a perennial crop with considerable latitude for manipulation of cultural practices.
5. Alfalfa fields contain a wide variety of insects and other organisms, many of which are highly beneficial.
6. Insects found in alfalfa fields commonly interact with adjacent crops.
7. There are many cultivars available with various degrees and types of pest resistance.
8. Pesticide residues must be minimal.

Features discouraging IPM:
1. Alfalfa has been commonly considered as a low value crop; while this is slowly changing, the reputation persists.
2. In many parts of the country alfalfa is a rotation crop or an ingredient in pasture mixes. When pests move in, growers switch to clovers or other crops.
3. Forage alfalfa in many areas is not a high pesticide-use crop, therefore the volume of pesticide use nationally will not be changed greatly even if all alfalfa pesticides were discontinued.

Before any integrated pest management program can function, there must be extensive data accumulated. In all programs there is constant interaction between the crop, the pests, the various pest-limiting factors, and economics. Basic to all of these factors are such things as long-range climate, short-term weather and soil types. If these factors can be measured, they can be used as predictive tools.

Our first need in the alfalfa pest management program is to recognize the basic factors limiting yields. These include many things in addition to pests, but in this discussion we will center on the pest problems. Each pest has many factors restricting its unlimited development. They are often referred to as natural controlling factors and include parasites, climate, plant resistances, etc. In working with forage alfalfa in northern Utah we decided the key insect pest was the alfalfa weevil, accounting for about $5 million loss in the state annually. Common secondary insect problems were pea aphids, about 5 species of caterpillars, lygus, and grasshoppers. The primary limiting factors related to the alfalfa weevil are: parasites, especially *Bathyplectes curculionis*; several predators; the harvesting dates of alfalfa crops; the type of harvesting; climatic conditions, especially during winter; and insecticide treatments. Each of these limiting factors is inter-related with many other factors and, except for climate, most are subject to a certain amount of manipulation. We are continually measuring these factors and expanding their use for both predictive and strategy functions.

A second major need in our pest management program is to improve sampleing and monitoring methods. No single technique serves all necessary functions. We make use of the sweep net, insects per terminal, the D-Vac, and visual damage ratings. The sampling method must be changed and modified to meet the requirements of the crop. Methods are changed according to the stage of alfalfa growth and expected pests in a given season.

The third key ingredient to our pest management program is to evaluate various control strategies. This evaluation must be related to factors within the alfalfa fields and to those outside the fields. The impact of one control strategy interrelates with other control strategies and with economics. We must consider three general control methods related to the alfalfa weevil that can be manipulated by growers: cultural practices, chemical control and making more efficient use of beneficial insects. Experimentally we work with other concepts such as the introduction of new parasites, but they presently are not part of grower strategies.

The basic model used in our work is that developed by the NSF-EPA supported alfalfa ecosystem studies. This work was done through an interstate cooperative effort in which Utah was a major cooperating state. The alfalfa plant model and alfalfa weevil model work rather well for us in our predictions. Unfortunately, the project was discontinued before other insect pests could be worked into the model. In Utah, we are now working with the economists to include an economics model into the system.

Several states are now using pest management systems relating to alfalfa pests. Most of these are on forage alfalfa, but a major effort is being made also against pests of seed alfalfa. In Indiana they have established a network of computer centers with regional divisions. In each of these regions a record of degree days, progress toward harvesting dates and unusual problems are keyed into the computer. A grower or a pest

41

management scout checks for alfalfa insects using a predetermined sampling method. His counts are phoned into the regional terminal and within a few minutes the advisory message is returned.

In Utah, due to the vast number of different conditions in the mountain country, we have been relying more on samples obtained at 50 degree day intervals. The decisions are made based on data relating alfalfa weevil numbers to growth stages of the alfalfa.

Under Utah conditions we can commonly suppress alfalfa weevil populations sufficiently to avoid insecticides by using cultural practices. The most effective cultural practice is cutting about a week early. During most years, the first crop of alfalfa can be cut before the alfalfa weevil larvae reach their peak feeding potential. Larvae that are still in the third instar are killed and the crop is cut before substantial injury appears. Sometimes it is necessary to use a stubble spray following harvest, but often cutting alone is adequate. Chemical use and cultural practices are manipulated in such a way that predators and parasites are preserved.

We can keep pests below economic injury levels without pesticides in more than one half of the fields during most seasons in Utah by manipulating cultural practices. This compares to about 60% of the fields normally sprayed for alfalfa weevil control plus about 10% sprayed for other pests. Of the approximately 35% of the fields under pest management requiring insecticide treatments, about half can use stubble sprays which do little harm to beneficial insects and create almost no residue problems. This leaves only about 15% of the fields requiring either early season or preharvest insecticide treatments. The preharvest insecticide treatments concern us the most. They can control the alfalfa weevil effectively, but create several serious problems: 1) They are applied during the height of beneficial insect activity. 2) Because treatments are made only 2-3 weeks before harvest, it is easy to miscalculate dosages or timings and get into residue problems. 3) When ground sprayers are used, considerable physical damage is done to the crop. Through the establishment of new parasites, manipulating other cultural practices, and more precise prediction methods we hope to continue reducing the number of preharvest treatments. In the meantime we recommend either selective or shortlived insecticides whenever possible.

THE NEED FOR INTEGRATED
PEST MANAGEMENT RESEARCH IN RANGE

B. Austin Haws

Department of Biology,
Utah State University

I would like to take this opportunity to discuss the need for expanded interdisciplinary research in range. There are approximately 960,000 acres of rangeland in the United States. Eighty to ninety percent of some western states are range. Users of range, technicians and range scientists are being asked to produce 30% more meat and milk from range by 1980. The increased cost of feed and the energy costs associated with the stockyard industry are placing renewed interest in rangeland.

One indicator of the need for intensive range research is the publication of new range policies for the various agencies involved with the management of our country's grasslands. These policies completely ignore insects as components of range systems. For example, the complete elimination of grazing on certain government lands may have disasterous effects in view of basic ecological principles and past experience with the soil bank. The lack of economically viable pest management alternatives for range management is further evidence of the need for integrated pest management research on range grasses.

When I began working with range insects in 1971, I soon realized how far behind we were. Surveys indicate that some aspects of range research and practices are about 35 years behind those of some other major crops. Integrated range research is still in its infancy. Only isolated and fragmented projects exist around the country. There is much to do and I feel it is an exciting time for research in range pest management.

The impact of insects on range is poorly understood. With the exception of grasshoppers and a few other insects, little is known about the basic biology and life cycles of range insects. Insects are important components of range ecosystems. Utah State University data show that a relatively low population of insects and their relatives consumed 2.8 AUMs (animal units per month) while livestock and wildlife ate 2.1 AUMs on a range in 1975. The basic data base is not yet available to develop a comprehensive pest management program in range.

The basic research needs in range include the identification of: the kinds and relative numbers of insects and their basic biologies found in range grasses; the major insect pests and beneficial range insects; the

economic thresholds for various range pests; the influence of all the
vital factors of their environment; the relation of the vertebrates
and invertebrates to various plant communities and to each other; the
relation of the lifecycles of the many range grasses to the animals, other
plants, climate, soil, etc.; the possibilities of breeding resistant grass
varieties; and the relationships of grazing, methods of planting, etc., to
the development of insect and weed pests.

Many of the fundamental principles of range management have been developed without a full complement of interdisciplinary research. In
fact, some range policies, recommendations and practices appear to promote insect development, serious insect outbreaks and unmeasured losses
in our ranges.

One management practice used in Utah is to remove sagebrush by spraying
with herbicides or burning. While this increases the available moisture
and nutrients for grasses it also reduces the diversity of habitat for
predators and parasites. High incidence of pest outbreaks have occurred
in some of these areas.

Another management practice to increase the amount of rangeland is chaining (removing) pinon pines and juniper. One chaining technique involves
chaining the pinon and juniper, leaving the debris in place, and planting seed among the remaining native grasses.   This technique provides
a diversity of plant species and habitat. Another technique involves
chaining the trees, removing them, and planting an introduced grass monoculture. Ken Ostlie, one of our students, has found that the black grass
bug, *Labops*, which is becoming a serious range pest in Utah, is 60 to 100
times more abundant in grass monocultures than in areas where the "debris
in place technique" is used. By reducing the diversity of habitat for
beneficial predators and parasites, serious infestations of range pests
can occur.

Some methods of grazing may also favor insect development. Most of the
eggs of certain insects are in the lower part of the plant and remain in
the field after grazing. If not removed by grazing the eggs are ready
to develop to their full biotic potential. The amount of plant litter in
ranges and the amount removed by grazing seem to be related to the number
of injurious insects present. Fields we have studied which have been
cleanly grazed in the fall or winter, or have been thoroughly burned,
usually have few injurious insects. Some insect eggs appear not to survive the trip through the digestive system of certain animals. Removing
these insect eggs by grazing may help explain why the grazing systems
of some ranchers are resulting in an increased weight gain of their stock.

The cooperative efforts of ranchers, range specialists from state and
federal agencies and Utah State University have begun to identify and try
to solve some problems of grass production. In 1971 USU initiated its
range research and at present we have a 10 man team. Our interdisciplinary
range research team is exploring possible pest management alternatives.
Cultural and managment practices, as well as chemical contol, offer promising and economical possibilities for reducing losses due to range insects.

Examples include: 1) proper planting methods, such as seeding in hetero-cultures instead of monocultures; 2) periodic grazing by single or combinations of animals, following the growing season, to reduce over-wintering insect populations; 3) use of grasses resistant to insect and nematode feeding, and diseases; 4) burning to remove plant residues and eggs; 5) protection and propagation of beneficial insects.

Planting resistant varieties of grasses shows great promise. Many of the grasses that have been introduced were selected on the basis of moisture adaptability and palatability and have not been screened for protection against major pests. Currently, little is being done on breeding grass resistance but in the few tests we have conducted, it looks like there is great resistance material available. In Utah there are at least 300 species of range insects many of which are beneficial. There are encouraging possibilities for biological control in range.

One important means to developing sound range management is to develop interdisciplinary curricula in our educational institutions. Students don't have the foundations to adequately approach the problem from an integrated perspective. We have to give students a broader understanding of ecosystems. Unfortunately, our educational system is still structured to provide students with only a small specialized piece of the entire picture.

As with other new approaches, one problem we face is the legitimacy of interdisciplinary range research. Tackling this involves getting the right information to all government levels in order to get the necessary support for this type of research. There has to be good communication between researchers and state and federal agencies to avoid duplication of effort and to combine our resources in solving problems of mutual interest.

Although much remains to be done in basic range research and the develop-ment of control alternatives, we are not starting from zero. Many princi-ples and procedures developed in other crop studies, such as alfalfa, will be applicable but the specific details related to grasses will need to be established. It will be exciting to study the basic biologies of range insects to build a panorama of the wildlife and plants, and to put them all together into an integrated system.

Someone said a good idea doesn't care who has it. We hope the agricultural research capability of the country will pick up on range IPM as it has on other crops. It was said recently in a meeting in Salt Lake City that it seems too often we don't have the money to do things right the first time, but we usually seem to find the money to do things over again. We hope there is a way to get enough financial support to develop pest management correctly for range grasses the first time.

PART THREE
CONTROL STRATEGIES

BIOLOGICAL CONTROL
BY NATURAL ENEMIES

Robert van den Bosch

Division of Biological Control,
University of California

BIOLOGICAL CONTROL is simply the regulation of plant and animal numbers
by natural enemies. As such it is a natural phenomenon, one of a
spectrum of physical and biotic forces which collectively maintain through
a balance of nature, the mechanism of natural control. Natural control
is vital to life on earth, for it is the phenomenon that permits the
co-existence of the planet's myriad species. Biological control, then,
as a major factor in the balance of nature, is of immense importance to
us .

In assessing the phenomenon of biological control, it can be viewed in
three major aspects: 1) Its naturalistic dimension, 2) the classic importation of natural enemies, 3) the preservation and augmentation of
natural enemies.

This discussion will be restricted essentially to an overview of the
biological control of pest insects and weeds, and will emphasize the
role of predators and parasites in biological control with only passing
mention being made of pathogens. Pest pathology (including microbial
control) is an immense subject in itself and cannot be given justice in
such a brief discussion as this.

Again, it is emphasized that biological control is an extremely important
natural phenomenon that is constantly going on all around us. It is
effected by a stupendous range of natural enemies against a vast array
of host (victim) species. This can perhaps be best visualized by considering the host-natural enemy interaction in its two extremes; on the
one hand, predation by the sperm whale on pelagic squid, on the other,
victimization of a bacterium by a bacteriaphage. Between these extremes
lie the countless predator-prey, parasite-host, pathogen-host interrelationships of nature.

It is perhaps best at this point to define a few terms that are commonly
used in a discussion of biological control. These terms are:

parasite: a small organism that lives in or on a larger host organism.
parasitoid: a parasitic insect that lives in or on, and eventually
           kills, a larger host insect (or other arthropod).
pathogen: a microorganism that lives and feeds (parasitically) on
           or in a larger host organism, and therby causes injury
           to it.
predator: an organism that feeds upon other species (animal or
           plant) that are either smaller or weaker than itself, or
           (in the case of plants) lack mechanisms of resistance
           or tolerance to it.

NATURALLY OCCURRING BIOLOGICAL CONTROL

In pest control evolution the past third of this century is known as
the synthetic organic pesticide era. The record shows that during
this period there was a massive increase in pesticide use, to the extent
that in many of the world's centers of agricultural (and other resource)
production, as well as in disease vector control, the chemical control
strategy has prevailed. But this strategy is falling short of expecta-
tions, and there is an increasing trend to integrated control (integrated
pest management). One of the key factors that has crippled the chemical
control strategy has been its adverse effects on naturally-occurring
biological control. This has led to a global syndrome, termed by some
the Pesticide Treadmill, which is characterized by frequent target pest
resurgences, secondary pest outbreaks, and pest resistance to pesticides.
As a result it is questionable whether we have made gains against the
pests since the mid 1940's, and in fact with insects there is evidence
that we have lost ground.

Despite the troubles associated with the chemical control strategy there
are some positive aspects to the debacle. One is the realization that
biological control as a ubiquitous natural force is an extremely important asset to us in our never-ending competition with pest organisms,
and that it can be built into, and indeed must be a pillar of the emerging
integrated control strategy.

All ecosystems, even crop monocultures, contain food chains and food webs
in which carnivorous and herbivorous species (predators, parasites and
often pathogens too) restrain or completely suppress pest or potential
pest species. Common sense dictates that we take full advantage of
this bonus mortality in our pest control strategy, and this is what is
being increasingly done under integrated control. Every integrated
control program of which I am aware has had a major biological control
component. This component occurs as naturally-occurring biotic mortality.
In this connection there is no question in my mind that in our future
employment of biological control the greatest emphasis will be on the
utilization of naturally-occurring predators, parasites, and pathogens.
In fact, as emphasis on integrated control research has increased, there
has been an immense amount of study on naturally-occurring biological
control. It is impossible to adequately treat these studies in this
short presentation; however, in an attempt to give some impression of
the range, diversity and importance of naturally-occurring biological
control in pest management, I have tabulated a number of successful integrated pest management programs, and have noted in a very superficial
way, the kinds of natural enemies involved (See Table 1).

CLASSIC NATURAL ENEMY INTRODUCTION (Classic biological control)

Classic biological control is largely directed against accidental invaders
of new areas which attain epidemic abundance because of their escape from
the natural enemies that restrain them in their native habitats. There
are some rare exceptions to this, but the overwhelming number of successes

Program	Locality	Type of natural enemy
Cotton	Peru	generalist predators and parasites
	California--USA (San Joaquin Valley)	generalist predators
Alfalfa	California, USA	predators, parasites, pathogens
Apple	Washington State, USA	predatory mites
	Michigan, USA	predatory mites
	Pennsylvania, USA	spider mite feeding ladybeetles
	Nova Scotia (Canada)	predators and parasites
Pear	California, USA	predators
Soybean	Midwest and Southeast USA	predators, parasites, pathogens
Wheat	Chile	predators, parasites, pathogens
Oil Palm	Malaysia	predators, parasites, pathogens
Tea	Sri Lanka	parasites
Citrus	Israel	predators, parasites
Highway vegetation	California, USA	predators, parasites
Mosquitoes	California, USA (Marin County)	predators

Table 1   Examples of integrated control programs with important biological control components.

in classic natural enemy introduction have involved the re-association
of invader pests with their adapted natural enemies obtained from the
areas of indigeneity.

Classic biological control involves three basic steps: (1) identification
of the pest's native home (this, of course, involves correct identification of the pest itself), (2) a search in the native area for the pest
and its natural enemies, (3) shipment of the natural enemies to the
invaded area and after appropriate quarantine processing and biological

51

testing, mass production, and colonization of these natural enemies in the field.

As was earlier indicated, the goal of this process is to re-establish the host-natural enemy relationship, and in doing so to lower the pest's long-term population level so that the species is reduced in severity or entirely eliminated as a pest. It is important to note that imported natural enemies do not eradicate their hosts. Instead they drive the host numbers down to and maintain them at reduced average densities. If the average is low enough the species remains below the economic, aesthetic or public health threshold and is no longer a pest. The question often arises,"What happens to the natural enemy when its host becomes scarce? Does it then attack other animals and crops and become a pest?" The answer, of course, is "no", because the natural enemy density is dependent on and most often specifically locked to the host pest. Thus it wanes in abundance as its host's numbers diminish, and becomes rare too, but with the capacity to regulate its host at a very low population level.

Worldwide, natural enemy introductions have resulted in substantial to complete control of slightly over 100 pest insects and about 20 weed species. This may appear to be a very poor record when measured against the earth's thousands of pest insect and weed species, but it must be noted that control by imported natural enemies is permanent. In this sense it is one of our most successfully employed pest control tactics. The natural enemy introduction tactic does have its limitations, particularly in that it has little potential against native pests (i.e., insects and plants that become pests in their native habitats). And even against exotic pests, it is not always possible to effect satisfactory biological control because of ecological, biological, administrative, technical, and logistical limitations or encumberances. Furthermore, the tactic has also been grossly under-exploited as evidenced by the fact that it has been utilized against only about 225 of the world's 10,000 or more pest insects; in many of these cases the programs were poorly conceived and conducted. Nevertheless, classic natural enemy importation has been one of our most successful tactics in effecting permanent pest control, and it has considerable potential for additional success.

BIOLOGICAL CONTROL OF WEEDS

The great majority of classic biological control successes have involved insects. However, the tactic has been successful against weeds too. In principle there is little difference between biological control of insects and weeds: (1) Both involve natural enemies; e.g. weeds (predators and pathogens), insects (predators, parasites and pathogens). (2) Successes with imported natural enemies have been overwhelmingly against alien (exotic) pests. (3) Both techniques offer safe, effective, long term control at low cost. (4) Control does not result in pest eradication.

But there are important differences between the two practices. With weeds, there is an absolute necessity that the natural enemy be highly specific, preferably monophagous[1] and at most very narrowly oligophagous.[2] There can be absolutely no possibility that the introduced enemy will develop an affinity for some cultivated or otherwise highly valued plants. On the other hand with entomophagous insects, oligophagy or even polyphagy may sometimes be advantageous, and there is certainly no stipulation that an imported parasite or predator be narrowly specific. The basic concern is merely that no beneficial insect (e.g. the honey bee, some native pollinators, lady beetles, lacewings, etc.) be adversely affected, or that hyperparasites be imported.

The need for specificity in weed feeding insects places an inordinate burden of responsibility on everyone involved in the importation process, particularly those who do the testing. In practice the candidate insect (or pathogen) is first intensively tested overseas in the collecting area. In this connection, all agencies which are seriously involved in weed biological control, maintain overseas laboratories (i.e., United States Department of Agriculture, Commonwealth Institute of Biological Control , Australia, and University of California). Testing involves wild plants related and unrelated to the weedy target and a wide range of related and unrelated commercial plants. The overseas testing establishes whether a species will be passed on to the domestic quarantine facility for additional intensive study and testing. Then, even after this second set of screenings is completed, the data are reviewed by a special committee of experts who make the final decision as to whether the natural enemy shall be cleared for field release.

The suppression of weedy plants by imported natural enemies differs somewhat from the suppression of pest insects. With insects, suppression usually results directly from premature mortality produced by the natural enemy. But with weeds the role of the natural enemy is more complex. (1) It may simply kill the plants. Here the timing of attack (e.g., at a time of nutritional stress) may be as important as massive defoliation. (2) It may act as an intermediary for some pathogen (e.g., as occurred with prickly pear in Australia). (3) It may destroy the reproductive capacity of the weed (i.e., seed feeders). (4) It may simply impair the weed's competitive capacity, so that it is displaced by more valuable species.

But despite the technical and subtle mechanical differences just described, biological control of insects and weeds operates under the same broad principles. At the heart of the matter is the phenomenon of reciprocal density dependence.

Biological control of weeds can be considered in the same three basic aspects as biological control of insects, or other animal species: (1) the naturalistic, that is, the broad phenomenon of plant species under regulation by natural enemies, (2) the classic introduction of natural enemies as discussed above, and (3) the augmentation and conservation of natural enemies. However, the latter is less applicable than with insects. In fact, outside of the use of pathogenic spores as exemplified in northern joint vetch control in rice, it is difficult to cite an augmentative practice utilized in weed biological control, unless it is the

[1]host range a single species. [2]restricted to a few related species

use of selective pesticides to protect a natural enemy, or re-inoculation of an area with an enemy agent.

Several factors bear on the consideration of a weed's suitability for biological control and here again, in part, there are differences from insect control. These factors are:

1. Whether the plant is native or introduced.
2. Whether the weed has close relatives that are of economic or aesthetic importance.
3. Whether the weed occurs in a disturbed (short crop) or stable situation.
4. The viewpoint of various groups as to the plant's harmfulness or value.

In other words, what is one man's weed is another man's resource: e.g., Johnson grass has some forage value; saltcedar is considered a weed by growers, but to fish and game people and hunters, it is a good thing because it serves as a nesting place for doves; yellowstar thistle is considered a pest by stockmen but an asset by fish and game people and apiarists.

These conflicts should be kept in mind. In working out the cost-benefit balance of a program it is important to remember that biological control of weeds doesn't eliminate (eradicate) the plants, and so it is possible to have matters both ways. Thus, conceivably, under biological control the virtues of a plant can be retained while its disadvantages are suppressed.

Biological control of weeds cannot simply result in the exchange of one weedy species for another. Thus, a program might have to entail introductions against two or more weeds if true benefit is to be realized.

Insects have been the major natural enemies used against weeds, but other organisms have been successfully employed. For example, fish have been used against aquatic weeds and geese against certain terrestrial weeds. Several pathogens have been used for weed control, as have mites and nematodes.

PRESERVATION AND AUGMENTATION OF NATURAL ENEMIES

Modern pest management as it shifts to the integrated control strategy increasingly incorporates and optimizes biological control. The addition of new natural enemies into crop or other resource systems by the classic importation of exotic natural enemies is one way in which this is done. But it is important as well to preserve or augment natural enemies that already exist in a resource environment, and this can be done in a number of ways. One way to do this is by periodic innoculative or inundative colonization of insectary reared natural enemies in the field. In the past there has been considerable question as to the usefulness of this technique. Mass release of the egg parasite *Trichogramma* in such crops as sugarcane, cotton, corn and apple has been especially suspect. On the

other hand, very successful programs of mass release of a variety of
natural enemies against citrus pests by California's Filmore Citrus
Protective District, releases of several kinds of natural enemies against
glasshouse pests in Europe, and release of parasites against the Mexican
bean beetle in soybean in the Southeast USA indicate that the periodic
colonization tactic can be effective and economical. Obviously the
matter needs further study and development.

A variety of cultural manipulations have been employed or proposed as
means for preserving or augmenting natural enemies. For example, nectar
plants have been used to provide a food source for adult parasites. Strip
harvesting of alfalfa has been effectively used in California to enhance
biological control of aphids and lepidopterous larvae. Also in California,
cover crops have been used in almond groves and vineyards to favor predators of spider mites. Again, in California, it has been shown that
planting of blackberry (*Rubus* sp.) adjacent to vineyards increases the
effectiveness of the parasite of the grape leafhopper. In this case the
grape leafhopper passes winter in diapause, but its parasite does not
and therefore must find an alternative host to survive. This host is
a species of leafhopper that occurs on blackberry. Thus, by planting
blackberry near vineyards, a population of the grape leafhopper parasite
is maintained and is available for early invasion of the vineyard each
spring.

Currently, there is considerable interest in nutritional augmentation of
natural enemies, and in the use of kairomones (cue chemicals), to enhance
their effectiveness. Results of limited testings have been very promising. But these aspects of natural enemy manipulation are still very
much in the research phase and are not now claimed as being effective
practical tools.

Without question, the most important area of natural enemy preservation-
augmentation lies in the selective, discriminate use of chemical pesticides. Of the many integrated control programs now in effect worldwide,
selective use of pesticides has been an important element in virtually
every one. As the integrated control strategy comes into wider implementation, the selective use of pesticides will almost certainly become
more widely, indeed almost routinely, used in pest management practice.

THE FUTURE OF BIOLOGICAL CONTROL

Biological control has a future, because as a natural phenomenon it will
continue to act upon pests whether we recognize this role or not. But,
of course, since we are increasingly recognizing it and incorporating
it into our pest management strategy, biological control will inevitably
play a more important role in pest control as time passes. In all probabil-
ity, naturally-occurring biological control will recieve our major
attention, but classic natural enemy importation will also receive increased emphasis. As our knowledge of, and techniques in, natural enemy
preservation and augmentation increase in depth and sophistication, biological control will become a much more effective pest management tactic.

DISCUSSION

QUESTION: *Have natural enemies from closely related pest species been used in biological control programs?*

VAN DEN BOSCH: *Yes, but first we go the easy route by looking for control agents associated with the pest species. There have been several cases where natural enemies from related species have been introduced and established in the field. The coconut moth from one of the Pacific Islands is one example where a parasite was obtained from another species. However, the overwhelming number of cases have used parasites obtained from the pest species in its native habitat.*

QUESTION: *Are entomologists looking at the possibilities of breeding insecticide resistance in parasites and predators?*

VAN DEN BOSCH: *Marjorie Hoy, who is now at Berkeley, is working on deliberately developing and selecting strains of predatory mites in the laboratory for use in integrated control programs in heavy pesticide-use systems. Stan Hoyt in Washington state discovered that Metaseiulus occidentalis had a tolerance for organophosphate insecticides. This was a real breakthrough in the spider mite IPM program on apple because Stan was able to incorporate that tolerance into the program.*

COMMENT: *I think it is wonderful that tests are being made to breed or select predators and parasites that are resistant to chemicals used in certain systems, but I think one should not let that obscure one's view of the total system. It is never simple in a biological system and it isn't usually a matter of just one parasite or predator. Too narrow a focus obscures the view that what you really want is a lot of natural enemies in the control zone.*

VAN DEN BOSCH: *Yes, maybe if we understood the entire system, we would find out that some of these problems would solve themselves. I have recently reviewed some data on forest entomology dealing with the tussock moth and the spruce budworm, pests of massive economic impact and tremendous public concern. You begin to understand why the tussock moth is becoming a more important pest; it is because of the way we are harvesting the forests. Under this program the forest is shifting over to a fir type, a monoculture, and under this situation the tussock moth population goes wild. You can go in and spray but, unfortunately, we have developed a nursery for the tussock moth. In a sense this is what has happened with the balsam fir and the spruce budworm in the northeast. Here we have protected over-mature trees and turned the forest into a perennially susceptible environment. We have to look at these systems in their entirety. We have to look at the resource, its total relationship with the environment in which it grows, and utilize that environment in every way we can to help us get maximum yield or quality out of the resource. Finally, if necessary, we have to use our artificial means, whether biological, chemical, genetic or cultural, in a very intelligent way.*

REFERENCES

DeBach, P. 1974. Biological Control by Natural Enemies. Cambridge University Press, 321 pp.

DeBach, P. (Ed.). 1964. Biological Control of Insect Pests and Weeds. Chapman and Hall, London, 844 pp.

Franz, J.M., and A. Krieg. 1972. Biologische Schädlings-bekämpfung. Pavey, Berline, 208 pp.

Hagen, K.S. and J.M. Franz. 1973. A history of biological control. In "A History of Entomology." (R.F. Smith, T.E. Mittler and C.N. Smith, Eds.), pp. 433-476. Annual Reviews Inc., Palo Alto, Ca. 517 pp.

Huffaker, C.B. (Ed.). 1971. Biological Control. Plenum Press, New York and London , 511 pp.

Huffaker, C.B., and P.S. Messenger. 1977. Theory and Practice of Biological Control. Academic Press, New York, San Francisco, London, 788 pp.

van den Bosch, R. 1971. Biological control of insects. *Ann. Rev. Ecol. Syst. 2:45-66.*

van den Bosch, R., and P.S. Messenger. 1973. Biological Control. Intext, New York, 180 pp.

CULTURAL CONTROL

Theo Watson

Department of Entomology,
University of Arizona

INTRODUCTION

The concept of integrated pest management (IPM) has gained recognition
and acceptance only in recent years. However, varying levels of pest
management have been practiced on a localized basis for a long time. The
methods available for combating pests have increased in numbers along with
the evolution of intensified agriculture in this country. Today, a sufficient number of control tactics are available to permit a more flexible
and satisfactory approach to pest control. This can be accomplished by
integrating these options, each of which imposes an additional repressive
effect upon pest populations, into a compatible and harmonius system. Among
these components of an IPM system is cultural control, the topic for discussion in this paper.

The term cultural control is so broad and all inclusive that a definition(s)
is germane at this time in order to unify our thinking on this subject.
Watson et al. (1976) have defined cultural control as the use of farming
or cultural practices associated with crop production to make the environment less favorable for survival, growth, or reproduction of pest species.
In addition to the direct repressive effects of cultural practices on
insect (or other) pests this definition encompasses the effects of other
control methods such as biological control as they are influenced by environmental change. For example, since biological control agents such as
predators and parasites are a part of the environment, any practice that
alters the environment to the enhancement of beneficial species would result in an additional repressive effect upon pest species. The National
Academy of Sciences (Anonymous 1969) publication on Insect-Pest Management
and Control states that the principle involved in the cultural control of
insect pests is purposeful manipulation of the environment to make it less
favorable, thereby exerting economic control of the pests or at least reducing their rates of increase and damage.

In a more natural setting such as our forests or wildlife preserves, the
terminology generally used is habitat management or environmental management. In a sense, this is analogous to cultural control in our agroecosystems. Komarek (1969) defined environmental management in its
broadest sense as the intelligent control and direction of the factors
affecting biological organisms. He further points out that the foundation for such direction is the study of the relationships of living

59

things to their environment and to each other - ecology. As with the biological control method, cultural control rests heavily upon ecological principles.

This method of control, or better still, management, is aimed primarily at preventing pest damage. Practices generally must be employed well in advance of the expected problem to achieve that effect. This requires an intimate knowledge of the biology and ecology of the pests involved, including their relationship to their host plants and to other biological organisms in the system.

The goal of cultural control, then, is to reduce pest populations to levels sufficiently low to prevent economic damage by the use of appropriate cultural practices. The practices, however, must be compatible with other aspects of the ecosystem such as optimum crop production. They should provide no benefit to other pest species and at the same time little or no detriment to beneficial species. This goal is definitely compatible with the modern philosophy of pest management.

The cultural control method possesses certain advantages as well as disadvantages over other control methods. Among the advantages are economy and non-disruptive effects. In many instances there is no additional outlay for equipment or operations. The effect can be achieved merely by altering the timing or procedure of an operation which would have been done anyway. Additionally, most cultural practices can be performed without the detrimental side-effects on beneficial species that generally follow the use of insecticides.

The major disadvantages of the cultural method are: practices need to be performed long in advance of the problem and the measures do not always provide complete control. Both of these, particularly the first, point to the need for a thorough knowledge of the insect's life and seasonal history and its habits, including host and habitat preferences. These disadvantages are gradually losing significance, however, as more knowledgeable pest management specialists are becoming directly involved in pest management programs on a continuing basis. They assist the grower in long-term planning aimed at preventing pest problems in addition to seeking immediate solutions to already-existing ones.

There are certain key considerations that must be taken into account in any decision to utilize cultural practices as a means of coping with pest problems or even to utilize them as a component in an insect pest management system. These key factors are: adequate knowledge of the biology and ecology of the pest, diversification of the cropping system, and availability of alternative control methods.

In general, a thorough knowledge of the pest's life and seasonal histories and its behavioral characteristics is necessary to effectively employ cultural practices for its control. The more that is known about a species, the greater is the likelihood that a weak link in the pest's biology can be utilized to develop cultural control techniques.

Diversification of the agroecosystems will determine to a large extent the options available for designing cultural control strategies based on the pest's biology and ecology. For example, because of the crops grown in the system, rotations may be ineffective due to the absence of the crop necessary to break the seasonal cycle of the pest. Alternative control measures, for example, chemical control, may make cultural control less appealing and perhaps less profitable for the short term. However, in the long term cultural control may be far more advantageous when considering the side effects that generally follow chemical control alone. In such a decision the grower must weigh the costs/benefits as a result of using the different methods against expected long-term effects.

Relative to a discussion of specific cultural practices, and as a matter of convenience, I will subdivide this topic into three broad categories: crop management, soil management, and water management. Various options available under each of these broad headings will be discussed.

CROP MANAGEMENT

Crop management provides an environmentally sound, effective and economical way to manage many of our major agricultural pests. Crop management practices can be employed to disrupt the life or seasonal histories of certain pests, to alter the microclimate creating a less favorable habitat, to permit escape from damaging infestations, or to utilize behavioral characteristics preventing damage to specific crops.

There are a number of ways of achieving one or more of the above through crop management. Among these are: crop rotation, planting or harvesting practices, crop diversification, and trap crops.

Crop Rotation

The basic principle involved with this cultural practice is the alternating of susceptible with non-susceptible crops to prevent the buildup of damaging infestation levels. Generally, pests having relatively long life cycles are the ones more effectively controlled by rotation. Control by this method does not require an exact knowledge of the life and seasonal history (Isely 1946) but does require an accurate knowledge of habits, particularly feeding habits.

The northern corn rootworm, *Diabrotica longicornis* (Say) is an excellent example of an insect that can be effectively controlled by rotation. This insect has a one-year life cycle (Metcalf et al. 1962) and the larvae must feed upon corn roots in order to reach maturity. Therefore, if any other crop follows corn the life cycle is broken and no damage occurs. Damage almost never occurs unless land has been planted to corn for at least two consecutive years.

Planting and Harvesting Practices

Planting practices can be used effectively in a number of ways to prevent damage from insect pests. The classic example involves a precise planting date for winter wheat to prevent infestation by the Hessian fly,

Mayetiola destructor (Say). This is based on an intimate knowledge of the insect's biology. Adult flies emerging in the fall live for only two to three days. Therefore, wheat sown late enough so that the plants have not come up before the fly emergence period will escape infestation. The proper date of seeding wheat to escape infestation by the fall generation of Hessian fly was worked out by entomologists in the experiment stations of all the principal wheat-growing states (Metcalf et al. 1962). As a result, fly-free dates have been established for various zones in the North Central states.

Another example of the importance of planting dates pertains to grain sorghum in Arizona. Sorghum is attacked by the sorghum midge, *Contarinia sorghicola* (Coquillett), which feeds upon the developing seed. Usually an economically damaging infestation does not occur until the fourth generation. Therefore, early-planted sorghum will reach maturity ahead of the 4th generation and escape injury. Another important cultural practice is to control Johnson grass, an important alternate host, near sorghum fields (Gerhardt and Moore 1962).

Harvest dates and practices are also important options in managing key pest problems. The harvest date of the first alfalfa cutting in the spring may be altered slightly to prevent serious damage by the alfalfa weevil, *Hypera postica* (Gyllenhal). Bishop et al. (1978) stated that harvesting the first crop of alfalfa before damage by the alfalfa weevil becomes severe prevents or delays the need for insecticide treatements in Idaho and Utah.

In the agroecosystem, the primary host of lygus bugs is alfalfa, on which populations increase to large numbers during the summer (Stern 1969). In most areas of the southwestern and western part of the cotton belt, both alfalfa and cotton are grown in large acreages. Lygus is a key pest of cotton and the problem on cotton is directly related to the practice of harvesting large blocks of alfalfa at one time and forcing the lygus over on the cotton. Stern et al. (1964) reported a strip-cutting system that provided two growth stages of alfalfa in the same field at all times. This reduced lygus emigration into cotton to the point where insecticide applications for lygus control were unnecessary.

Rakickas and Watson (1974) compared seasonal population fluctuations of adult and immature lygus and 7 predaceous arthropod groups, before and after cutting, in strip-cut alfalfa (Figure 1). In addition to the herding of lygus back and forth within the alfalfa field throughout the season, of particular significance was the maintenance of large predator populations for the entire season. Van den Bosch and Stern (1969) have reported on the effects of strip-cutting on other arthropods in alfalfa. Their results showed a general upward trend of predator populations in strip-cut fields as compared with a leveling off of populations in solid-cut fields. In addition, in the strip-cut fields the several predator species studied showed much less violent population fluctuations than they did in the solid-cut ones.

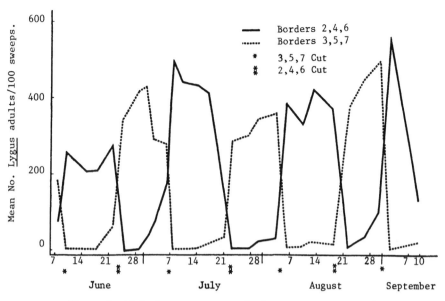

Figure 1 Adult Lygus movements in 2-stage strip-cut alfalfa.

Van den Bosch et al. (1967) have also reported on the effects of strip-cutting on the pea aphid, *Acyrthosiphon pisum* (Harris) and the parasite, *Aphidius smithi* Sharma and Roo. They showed that strip-cutting had a distinctly favorable effect on the parasite. In the strip-cut alfalfa populations of both species were maintained throughout the summer, permitting continuous contact between the parasite and its host, allowing for quick response to the aphids' upsurge in late summer.

Crop Diversification and Planting Patterns

Crop diversification creates conditions for a more complex interaction of organisms within agroecosystems, including pests and their natural enemies. These conditions are conducive to a more stable biological system because of additional checks and balances. In less complex systems, fewer species of both pests and beneficial insects exist but conditions are more favorable for wider amplitude in population fluctuation. In fact, this generally occurs.

Where diverse agriculture occurs, a conscious effort to arrange the crops within the system to take into account mutual pest problems may be used effectively to minimize or prevent pest problems on certain crops. For example, in Arizona late-season populations of salt-marsh caterpillar,

Estigmene acrea (Drury), may build up on cotton. Generally, it is so late that insecticidal control for the insect on cotton is unnecessary. However, fall lettuce planted adjacent to such a cotton field is vulnerable to the masses of migrating larvae as they leave the cotton. Some separation between such crops would alleviate this problem, even though other satisfactory control measures are available to protect the lettuce, e.g., a foil barrier around the lettuce field to prevent the migrating larvae from entering the field.

Bishop et al. (1978) stated that diversity resulting from the "dirty field technique"(allowing noncrop plants to survive rather than using clean culture) has a similar stabilizing effect on the agroecosystem. Diversity, however, may not be the most efficient and economical short-term method of crop production.

Trap Crops

Trap cropping is simply the planting of a more favorable host crop in the system to attract the pest and prevent it from attacking a more valuable crop. This requires a general knowledge of the life history of the pest and, more importantly, an intimate knowledge of host preferences of the pest.

In the South, prior to the availability of organic insecticides to control the bollworm, *Heliothis zea* (Boddie), on cotton, trap cropping of this pest was relatively common. This involved planting small acreages of corn in the vicinity of cotton fields so that silking corn was available for oviposition by the bollworm moth during the major fruiting period of cotton. For season-long protection, this required a succession of plantings at two-week intervals (Isley 1946). It should be pointed out, however, that unless properly implemented this can backfire by sustaining large pest populations that would go to cotton if silking corn were unavailable.

SOIL MANAGEMENT

Soil management may be useful in minimizing pest problems for those pests that pass at least part of their life cycle in the soil. This may also be closely allied with the cropping system and the rotation scheme. The sand wireworm, *Horistonotus uhlerii* Horn, is an example of a species whose importance as a pest depends more upon the type of soil in which it lives than any other factor (Isley 1946). Changes in the crops grown would have little direct effect as this pest is a general root feeder and would thrive on a wide variety of crops. However, a crop that would add a significant amount of humus to the soil, and thus increase its water holding capacity, would aid in control. Soil management is most effective for control of this pest where the land can be fallowed for one or more years or where crops are grown that require no cultivation and are not attractive to the beetles during the period of oviposition. This could be accomplished with such crops as oats, grapes or clovers.

WATER MANAGEMENT

Water management offers a great opportunity to adversely affect a variety of pest problems, ranging from agricultural pests to those directly affecting human welfare. From the agricultural standpoint this is particularly true in the arid regions of the world. Since most insects have a rather narrow range of tolerance for moisture, there is the opportunity to either reduce the moisture level below the tolerance level or to provide an excess to raise it above the tolerance level. Research has shown that the pink bollworm, *Pectinophora gossypiella* (Saunders), requires an optimum level of moisture for overwinter survival, above or below which results in greater mortality (Slosser and Watson 1972a). Probably one of the most striking examples of controlling a pest with water management, and one with which we are all familiar, is drainage of standing water to eliminate a mosquito problem.

There are many other techniques, or modifications of those already discussed, that are classified as cultural methods and can be used effectively in localized situations. Some of these are: sanitation, destruction of crop residue, flaming or burning and elimination of alternate hosts.

The use of resistant host plants should also be mentioned since it is sometimes classified as a cultural practice. Others, however, consider it of such importance that it is elevated to the status of other major control methods. Where available, the use of resistant varieties is one of the most convenient and effective methods of pest control. The development of alfalfa varieties resistant to the spotted alfalfa aphid, *Therioaphis maculata* (Buckton), provided a most satisfactory solution to a very serious pest problem.

INTEGRATION OF CULTURAL PRACTICES TO ENHANCE PEST MANAGEMENT

It is unfortunate that when we think of cultural control, we usually tend to rely upon a single practice to solve the problem. In reality, however, there may be a number of cultural practices, each of which provides a segment of control resulting in overall satisfactory suppression. This is the case with the pink bollworm on cotton in the Southwest and in Texas where essentially a combination of practices resulted in effective cultural control of this pest (Adkisson and Gains 1960, Noble 1969). Adkisson and Gaines (1960) listed the primary practices as follows:

1. Defoliate or desiccate the mature crop to cause all bolls to open at nearly the same time, expediting machine harvesting.
2. Harvest the crop as early and in as short a time as possible.
3. Shred stalks immediately following harvest.
4. Plow stalks under immediately, preventing regrowth of new fruting forms that might provide food for diapausing larvae.
5. Prepare land for planting of subsequent crop, including pre-plant irrigation in arid areas.
6. Plant new crops during a designated planting period that allows for maximum "suicidal emergence" of moths from overwintering larvae.

The combination of practices developed in Texas provided the growers with a control program for the pink bollworm so successful that insecticides are seldom needed (Adkisson 1972).

During the past several years research has been conducted in Arizona on cultural control of the pink bollworm. The primary objective has been to reduce overwintering populations of diapausing larvae to levels sufficiently low to prevent economic infestations during the subsequent growing season.

A large amount of data showing the effects of various cultural practices on overwinter survival and spring moth emergence has been accumulated (Watson and Larsen 1968, Watson et al. 1970, Slosser and Watson 1972b, Watson et al. 1974). These data have shown that with each additional practice imposed on cotton, such as discing following shredding, plowing following discing, etc., there is an additional reduction in paring moth emergence (Figure 2). With the long growing season and mild winter conditions that occur in Arizona and regardless of the type of cultural

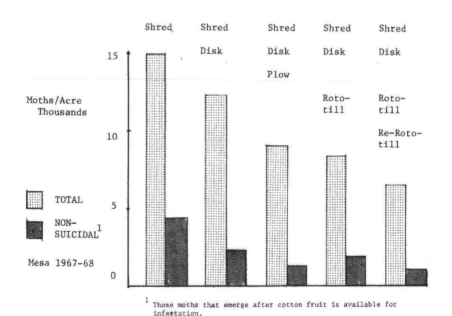

[1] Those moths that emerge after cotton fruit is available for infestation.

Figure 2   Effect of additive cultural practices on spring moth emergence.

practices used, overwinter survival and spring moth emergence are still too great to prevent economic infestations during the next growing season. The best results have been through shortening the growing season by earlier plowdown dates. These results showed that progressively higher moth populations emerged the following spring with each delay in plowdown date (Watson et al. 1974) (Figure 3). Essentially the same results can be achieved by earlier crop maturity without regard to when the cotton is harvested and plant debris plowed under. This utilizes knowledge of the biological phenomenon, diapause, to prevent build-up of overwintering populations. In Arizona the incidence of diapause in the pink bollworm is low until the last week of September, at which time it increases rapidly (Watson et al., 1976). Therefore, cotton that has attained the stage of maturity, where most of the green bolls are three weeks old by the last week of September, is incapable of building large overwintering populations.

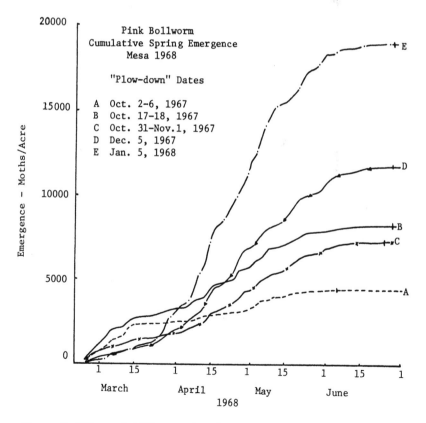

Figure 3 Effect of different plow-down dates on spring moth emergence.

67

Previously, the general assumption has been that early crop termination would result in lower yields. However, research conducted at the Yuma Valley Agricultural Experiment Station of the University of Arizona has shown little, if any, decrease in cotton yields when irrigation was terminated from August 1 to 15 as compared to termination in late September (Jackson and Carasso, unpublished data).

The objective in shortening the growing season is to preclude the development of large overwintering populations and thus eliminate or reduce the need for insecticidal control of the pink bollworm during the next growing season.

In 1971, a 3-year experiment was initiated at the Yuma Station to determine the effects of various irrigation termination dates and levels of pink bollworm control on cotton yields and overwintering pink bollworm populations as measured by spring moth emergence.

This research demonstrated conclusively that shortening the growing season by proper managment of irrigation water was the key practice in establishing an effective cultural control program against the pink bollworm. Additional cultural practices such as discing and deep plowing should be used to complement the effects of shortening the growing season. The earliest termination date, mid-July, proved to be too early and significant yield losses occurred. However, yields obtained from the three remaining termination dates, ranging from about July 29 to September 12 were comparable (Table 1). Of major signifcance was the effect of crop termination on moth emergence the following spring.

Termination Dates			Mean lbs. Seed Cotton/Plot[1] and Stat. Sig. (.05)[2]		
1971	1972	1973	1971	1972	1973
July 15	July 17	July 10	37.5a	33.4 n.s.[3]	53.4a
July 29	Aug. 4	July 31	57.3b	38.4	67.2b
Aug. 16	Aug. 23	Aug. 21	52.5b	40.5	63.3b
Sept. 3	Sept. 7	Sept. 12	56.5b	40.0	73.2b

[1] Harvested plot yields were converted to a standardized plot size of 0.02 acres for the three years.

[2] Yields in same column followed by common letters are not significantly different.

[3] n.s. means not significant.

Table 1   Mean plot yields for various irrigation-termination treatments. Yuma, Arizona.

The early August termination date resulted in low spring moth emergence, comparable to that from the mid-July termination date, but with higher yields. From the last two termination dates high numbers of moths emerged the following spring, indicated that the longer growing season provided adequate food at the critical time when diapause induction occurred (Slosser and Watson 1972b, Watson et al., in press). An example of the effects of irrigation cut-off dates on spring moth emergence is shown in Table 2.

Irrigation Cut-Off Date	Mean No. Moths/Acre and Stat. Sig. (.05)[1]		
	Immediate Plowing	Delayed Plowing	Combined
July 17	875a	625a	750a
Aug. 4	1,375a	1,250a	1,313a
Aug. 23	5,958b	6,917b	6,438b
Sept. 7	13,917b	18,417c	16,167c

[1] Yields in same column followed by common letters are not significantly different.

Table 2   Effect of irrigation cut-off and post-harvest plowdown dates in 1972 on pink bollworm moth emergence in 1973 (Means of 3 treatment levels) Yuma, Arizona.

The utilization of production practices to shorten the growing season while maintaining yields resulted in a conservation of irrigation water, a decrease in insecticide usage, and a reduction in overwintering pink bollworm larvae. This, coupled with other post-harvest cultural practices such as stalk-shredding, plowing and irrigation, has the potential of relegating the pink bollworm to minor pest status. In fact, growers who have practiced early termination have had few pink bollworm problems in their cotton production.

To summarize, it should be emphasized that cultural control, like biological control and other pest management tactics, is a population lowering procedure. This may require a re-education of the grower so that he understands the concept of living with continuous pest populations but at levels sufficiently low that economic damage is unlikely. It will also require that the growers and their pest management advisors think in terms of long-range control rather than for convenience and short-term profits only.

REFERENCES

Adkisson, P.L. 1972. Use of cultural practices in insect pest manage-
ment pp. 37-50. In: Implementing Practical Pest Management Strategies
Proc. Nat. Ext. Insect Pest Mgmt. Workshop. Purdue University,
Lafayette, Ind. March 14-16, 1972. 647 pp.

Adkisson, P.L. and J.C. Gaines. 1960. Pink bollworm control as related
to the total cotton insect program of Central Texas. Tex. Agric.
Exp. Sta. Misc. Publ. 444. 7 pp.

Anonymous. 1969. Insect-Pest Management and Control. Nat. Acad. Sci.
Publ. 1965. 508 pp.

Bishop, G.W., D.W. Davis and T.F. Watson. 1978. Cultural practices in
pest management. In Environmental Improvement through Biological
Control and Pest Management. *Western Regional Bulletin (In press)*.

Gerhardt, P.D. and Leon Moore. 1962. Sorghum midge--a new pest in
Arizona. *Prog. Agric. in Ariz. Vol. XIV (1):12-13*.

Isely, Dwight. 1946. Methods of Insect Control (Part I). Burgess
Publ. Co. 134 pp.

Komarek, E.V., Jr. 1969. Environmental Management. Proc. Tall Timbers
Conf. on Ecol. Anim. Contr. by Habitat Mgmt. 1:3-11

Metcalf, C. L., W.P. Flint and R.L. Metcalf. 1962. Destructive and
Useful Insects. McGraw-Hill, New York 1087 pp.

Noble, L.W. 1969. Fifty years of reasearch on the pink bollworm in
the United States. USDA Agric. Handbook No. 357. 62 pp.

Rakickas, R.J. and T.F. Watson. 1974. Population trends of *Lygus*
spp. and selected predators in strip-cut alfalfa. *Environ. Entomol.
3:781-4*.

Slosser, J.E. and T.F. Watson. 1972a. Influence of irrigation on
overwinter survival of the pink bollworm. *Environ. Entomol. 1(5):
572-576*.

Slosser, J.E. and T.F. Watson. 1972b. Population growth of the pink
bollworm. Ariz. Agric. Exp. Sta. Tech. Bull. 195. 32 pp.

Stern, V.M. 1969. Interplanting alfalfa in cotton to control lygus
bugs and other insect pests. Proc. Tall Timbers Conf. Ecol. Anim.
Contol Habitat Mgmt. 1:55-69

Stern, V.M., R. van den Bosch and T.F. Leigh. 1964. Strip-cutting
alfalfa for lygus bug control. *Calif. Agric. 18(4):4-6*.

van den Bosch, R., C.F. Lagace and V.M. Stern. 1967. The interrelation-
ship of the aphid, *Acyrthosiphon pisum*, and its parasite, *Aphidius
smithi*, in a stable environment. *Ecology 48:993-1000*.

van den Bosch, R. and V.M. Stern. 1969. The effect of harvesting practices on the pink bollworm in Arizona. *J. Econ. Entomol. 61(4):1041-1044*.

Watson, T.F. and W.E. Larson. 1968. Effects of winter cultural practices on the pink bollworm in Arizona. *J. Econ. Entomol. 61(4):1041-44*.

Watson, T.F., W.E. Larsen, K.K. Barnes and D.G. Fullerton. 1970.
Value of stalk shredders in pink bollworm control. *J. Econ. Entomol.
63(4):1326-1328*.

Watson, T.F., K.K. Barnes, J.E. Slosser and D.G. Fullerton. 1974.
Influence of plowdown dates and cultural practices on spring moth
emergence of the pink bollworm. *J. Econ. Entomol. 67(2):207-210*.

Watson, T.F., Leon Moore, and G.W. Ware. 1976. Practical Insect Pest
Management. W.H. Freeman and Company. 196 pp.

Watson, T.F., F.M. Carasso, D.T. Langston, E.B. Jackson, and D.G. Fullerton.
197_. Pink bollworm control in relation to crop termination. *J.
Econ. Entomol. (In press)*.

PHEROMONES AS
THIRD GENERATION PESTICIDES

Everitt R. Mitchell

Science Education Administration,
U.S. Department of Agriculture

Many researchers are investigating the use of sex attractant pheromones for managing insect pest populations. Already these potent chemicals are widely used to monitor seasonal changes in populations and to forecast potential problem areas, locate pest infestations previously undetectable by more conventional methods, and to schedule insecticide applications. These materials also are being evaluated for direct control of several important pests by mass trapping or disruption of premating communication. This paper discusses some of the most promising developments along these lines.

DISRUPTION OF MATING

Sex pheromones may provide control of insect pests by disrupting mating and thereby reduce subsequent larval infestations. The mechanisms by which disruption are accomplished are not well understood; the term atmospheric permeation, which implies no specific mechanism, has generally been accepted as descriptive of this approach.

Field Crops

Several recent studies have demonstrated control of insect pests via air permeation. Shorey et al. (1974) treated a 4.8 ha. cotton field with hexalure-impregnated cotton string evaporators during the summer of 1972. Hexalure is an attractant for male pink bollworms, *Pectinophora gossypiella* (Saunders), but is not the sex pheromone. This test was repeated in 1973 with different types of hexalure evaporators and evaporator spacings. Inspection of immature cotton bolls at the time of highest potential damage (mid-August) indicated that numbers of pink bollworm larvae were reduced 83% to 93% compared with control fields. In another test, hexalure-treated fields and fields treated 4 to 8 times with carbaryl had approx. the same level of larval infestation in cotton bolls.

This paper reports the results of research only. Mention of a commercial or proprietary product or of a pesticide in this paper does not constitute a recommendation for use by the U.S. Department of Agriculture nor does it imply registration under FIFRA as amended.

Gossyplure, the true sex pheromone of the pink bollworm, was identified in 1973 and proved to be a much more effective mating disruptant than hexalure. Therefore, during the 1974 growing season, gossyplure was used to treat approximately 1600 ha of cotton in the Coachella Valley of California (Shorey et al. 1976). A total of 9 g of gossyplure/ha was distributed over the season, giving a release rate of 12/mg/24hr/ha. Through mid-August, the larval infestation in bolls was comparable with that observed during the 3 previous seasons in fields that received conventional insecticide treatments. Also, there was a 3- to 4-week delay in the onset of larval infestations in the bolls in 1974 as compared to previous years. However, conventional insecticides were required in some areas of the Valley to provide late-season control of the pink bollworm. The investigators concluded that the 40 m separation between glossyplure evaporators was probably marginally wide for providing effective control.

Gaston et al. (1977) treated 5-, 6-, and 12-ha cotton fields with gossyplure during the 1976 growing season. Pheromone was evaporated from hollow, 104-mm long, thermoplastic fibers (Conrel[R] fashioned into hoops of 1.5 revolutions). The hoops were attached to the cotton plants by hand at 3-week intervals (from mid-May through early September) on a 1 X 1 m grid throughout each field (five applications). The fields treated with gossyplure and nearby control fields treated as needed with conventional insecticides were then monitored weekly with gossyplure-baited traps; also, larval infestations in bolls were monitored weekly in each field. Monitor traps positioned in the fields treated with gossyplure captured 98% fewer pink bollworm males than pheromone-baited traps located in fields treated with insecticides. Moreover, there was a substantial reduction in the number of pink bollworm larvae in cotton bolls in the gossyplure-treated fields vs. the insecticide-treated fields. Results showed an average 9-fold reduction in insecticide applications per hectare in the pheromone-treated fields compared with applications in the fields treated solely with insecticides for pink bollworm control.

Large-scale field trials in which gossyplure was used as a mating disruptant for the pink bollworm also were conducted in cotton in Arizona by Conrel, FRL-Albany International, Norwood, Mass., during the 1976 growing season. Their results showed gossyplure to be an effective control for this pest when used in conjunction with conventional insecticides (Roger Kitterman, personal communication). Only approximately 15% of the fields in this test required any insecticide, and the number of applications of insecticide in the fields that did require treatment was reduced 50% to 60% from the number to protect the control field, which was treated conventionally. Many growers in Arizona are considering a total pheromone-insecticide program in 1978.

Marks (1976) reduced the level of mating of *Diparopsis castanea* Hemps (the "red bollworm" of cotton in Central and Southern Africa) by evaporating its sex pheromone (dicastalure) in a 0.2 ha field cage. Dicastalure at 21 and 42 g/ha produced average reductions in mating of 47% and 72%, respectively, for one month. An inhibitor of male sex attraction (E)-9-dodecen-1-ol acetate, applied at the rate of 37 g/ha reduced mating by 71%. He also showed that the degree of mating disruption in these tests was density-independent for moth populations of up to 2,200/ha.

The feasibility of using the air permeation technique for mating control of the corn earworm, *Heliothis zea* (Boddie), and fall armyworm, *Spodoptera frugiperda* (J.E. Smith), was demonstrated by Mitchell et al. (1974, 1975, 1976) with (Z)-9-tetradecen-1-ol formate (Z-9-TDF) or (Z)-11-hexadecenal (Z-11-HDAL) used for the corn earworm and (Z,E)-9,12-tetradecadien-1-ol acetate (ZETA) used for the fall armyworm. Z-9-TDF is a chemical of non-biological origin; Z-11-HDAL has been reported as a component of the corn earworm sex pheromone (Roelofs et al. 1974); and ZETA is a part of the sex pheromone of several *Spodoptera* spp., though it is not a pheromone of the fall armyworm (Mitchell and Doolittle 1976). In these tests, mating by corn earworm females was suppressed 85% in plots treated with Z-11-HDAL and 96% in plots treated with Z-9-TDF. In plots treated with ZETA, mating by fall armyworm females was reduced 88%.

The majority of the experiments conducted to date with semiochemicals used for pest control have concentrated on single key pests. However, it is manifest that modern pest management strategies be devised to attack the problems created by complexes of pests, often in complexes of crops. Where it is feasible, manipulation of the behavior of several coexistent insects must be incorporated into an overall strategy. The opportunity exists to demonstrate such management capability with behavior-modifying chemicals that can be formulated to provide a multi-chemical attack on mating behavior.

Mitchell (1975) and Mitchell et al. (1976, 1977) proposed and demonstrated the feasibility of using multi-chemical formulations to disrupt mating among coexisting pest insects, the corn earworm and fall armyworm. When Z-9-TDF and ZETA were evaporated in separate plots, matings by the corn earworm and fall armyworm were reduced 96 and 88% respectively. When Z-9-TDF and ZETA were evaporated simultaneously in the same plot, matings by corn earworm and fall armyworm females were reduced 87 and 92%, respectively; a clear indication of the compatability of these two chemicals.

On the basis of these encouraging results, a 4-year research program will be initiated in 1978 to develop the technology for suppressing populations of the corn earworm and fall armyworm in sweet corn. The use of mating disruptants against these pests should permit a significant reduction in the number of insecticide applications (18-24) now required for control of pest insects in this crop. Moreover, the knowledge gained from these tests should be widely applicable to a broad spectrum of insect pests of field and vegetable crops in areas where it may be desirable to use air permeation alone or in other areas where integrated pest management programs involve conventional pesticides, biocontrol agents, and good cultural practices.

Fruit Crops

Air permeation also is being investigated as a possible control method for several important insect pests of fruit crops. The "summer fruit tortrix", *Adoxophyes orana* F.V.R., is the chief insect pest in Dutch apple orchards. The pheromone of this species has been identified as a mixture of 2 positional isomers, (Z)-9- and (Z)-11-tetradecen-1-ol acetate (Meijer et al. 1973). The geometrical isomers (E)-9- and (E)-11-tetradecen-1-ol acetate were described as antipheromones (compounds which oppose normal action of a pheromone) for *A. orana* because they almost completely blocked attraction to traps baited with a mixture of pheromone and either one or both compounds (Minks and Voerman 1973). Later, Minks et al. (1976) treated a 0.2-ha plot of an apple orchard with a microcapsular formulation of *A. orana* antipheromones, (E)-9- and (E)-11-tetradecen-1-ol acetate in the proportion of 9:1, against the 2nd flight of this species during August and September 1974. Three spray applications, each containing 8 g active ingredient, were made at two week intervals beginning August 1. Pheromone-baited monitoring traps placed in the treated area captured approximately 94% fewer *A. orana* males than pheromone traps located in a commercially sprayed area in the same orchard. In addition, the number of larvae on fresh shoots before and after spraying showed a significant depression in the number of *A. orana* of the next generation in the pheromone-treated area compared with the area treated with insecticides. These results could not be duplicated in 1975 or 1976, possibly because of hot weather during the test periods (Minks 1976, personal communication).

The codling moth, *Laspeyresia pomonella* (L.), is the most destructive pest of apples and pears in most parts of the world. Mating control by permeation of the air with codlelure, (E,E)-8,10-dodecadien-1-ol, appears to offer a viable alternative to conventional insecticides for suppression of this pest. In field tests conducted in Yakima, Washington, during 1973, 1974, 1975 (Moffit 1975) and 1976 (Moffitt, personal communication), communication between the sexes and subsequent mating of the codling moth was disrupted to a very high degree by applying excess concentrations of pheromone as a spray to orchards. The codlelure used in these tests was formulated in microcapsules and applied by heliocopter. In one test, two consecutive applications of codlelure made 8 days apart resulted in greater than 95% reduction in codling moth males captured in pheromone traps located in the treated area compared with the catch in traps located in an area that was not treated with codlelure. Furthermore, fruit harvested from the codlelure-treated area had 93% less larval infestation than fruit harvested from the untreated control area.

Stored Products

In studies conducted in Florida, Sower and Whitmer (1977) were successful in reducing reproduction and the rate of population development in two stored-product moths, *Plodia interpunctella* (Hübner) and *Ephestia cautella* (Walker), by permeating the atmosphere of rooms containing quantities of loose-stored peanuts with their respective sex pheromones. This test established that at low population densities, damage by *P. interpunctella* (Indian meal moth) and *E. cautella* (almond moth) can be reduced by using the sex pheromone (Z,E)-9,12-tetradecadien-1-ol acetate.

The air permeation technique is also receiving serious consideration as a method of regulating some important forest pests. The Canadian Forest Service is exploring the possibility of distributing fulure (a 97:3 blend of (E)- and (Z)-11-tetradecen-1-ol), to suppress population growth of the eastern spruce budworm, *Choristoneura fumiferana* (Clemens) (Sanders, personal communication). In addition, G.E. Daterman and co-workers (personal communication) have obtained good efficacy data on disruption of pheromone communication in the Douglas-fir tussock moth, *Hemerocampa pseudotsugata* McDunnough, on small (1 ha) plots by using point-source releases. They plan to move toward evaluation of an "operational" control-release system in the near future; and they also plan to evaluate mating disruption as a possible control system for the "western pine shoot borer", *Eucosoma sonomana* Kearsott.

PHEROMONE TRAPS

The possibility of using pheromones as a survey tool has received wide attention. By relating daily changes in the number and distribution of *Spodoptera littoralis* Boisduval males captured in a network of pheromone traps scattered throughout Cyprus with the prevailing meteorological conditions, Campion et al. (1977) was able to show that this species was endemic to the island. Thus, there was no evidence to support the hypothesis that *S. littoralis* migrates to the island each year from neighboring countries on the mainland.

Spodoptera exempta (Walker) is an important pest of pastures and related field crops such as corn in East Africa. This species is highly migratory in habit, and changes in its distribution can be related to synoptic weather (Brown et al. 1969). The establishment of a network of light traps throughout the region has made it possible to develop and early warning system to alert farmers of incipient armyworm outbreaks.

The recent indentification of the pheromone of *S. exempta* (Beevor et al. 1975) allows a more comprehensive trapping system, particularly in the more remote parts of East Africa where electric power is unavailable. Field trials with pheromone traps indicate a close correlation between light and pheromone trap catches, so the two trapping systems can supplement each other. This is particularly important at periods of full moon when light trap catches are depressed, but moth captures in pheromone traps are unlikely to be affected.

For several species, pheromone traps have proved to be a reliable device for early detection of the adult population and for indication of possible infestation by larvae. Minks and deJong (1975) devised a method for scheduling sprays to control *A. orana* in Dutch apple orchards based on pheromone trap catches and temperature recordings. Prediction of egg hatch is based on observations during embryonic development of the eggs. A model has been developed for quick calculation of the stage of egg development as a percentage of total development. As soon as the cumulative percentages exceed 100, hatching can be expected, and advice to farmers to "spray now" can be issued.

The pheromone trap has greatly facilitated study of the population dynamics and control of the codling moth. Croft (1975) and Riedl et al. (1976) have developed an extensive model system that utilizes pheromone trap catch data taken in early season as reference points and couples them with a physiological time model that forecasts critical events (egg hatch, emergence, damage, etc.) for spray timing. Inputs needed to use the model are first trap catch and peak catch from the spring moth generation. When coupled with online weather inputs, this system provides output options including predictive statements of developments, maps of development, and forecasts of life stage changes for the two generations of moths occurring in 27 sites across the Michigan fruit belt.

Limiting crop injury by trapping insects with pheromones has generally been unsatisfactory, especially in situations involving moderate to high populations where only one sex of the species is affected. However, aggregation pheromones that attract both males and females to a common site appear to be more amenable to this approach.

The "ambrosia beetle", *Gnathotrichus sulcatus* (LeConte), is a pest of freshly sawn, unseasoned lumber in sawmills in the Pacific Northwest. The presence of these beetles in export lumber from the west coast of North America (Milligan 1970) has led to quarantine problems with importing countries that have extensive exotic forests. Also, because of the ability of the ambrosia beetle to complete its life cycle within sawn lumber (McLean and Borden 1975), there is a need for protecting freshly sawn lumber from this insect (McLean and Borden, in press).

The population aggregation pheromone for the ambrosia beetle, sulcatol (6-methy-5-hepten-2-01), was isolated, synthesized, and successfully field tested by Byrne et al. (1974). McLean and Borden (in press) developed a mass-trapping technique with sulcatol to suppress populations of ambrosia beetle in newly sawn lumber. Briefly, sulcatol-baited traps are strategically placed next to piles of attractive fresh slabbing that could be colonized by beetles not captured in the traps. At the end of attractive period, approx. four weeks, the slabs are removed and chipped, thus killing any beetles that attacked them. This system was successfully field-tested in a commercial sawmill on Vancouver Island, B.C., during 1975 (McLean and Borden, in press), and 1976 (J.A. McLean, personal communication). Although the system has not provided 100% control of the problem, industry has been sufficiently impressed with the results to continue the trap-out program on their own as a part of their quality control procedure (J.H. Borden, personal communication).

FORMULATION

Before insect sex attractant pheromones or other semiochemicals can be used effectively, they must be incorporated into a system that will give a constant, dependable level of chemical release and protect them from the degradative action of the weather. Dispensers of many types, including cotton dental wicks, plastic vials and caps, polyvinyl chloride rods, rubber bands, and rubber stoppers have long been used to dispense pheromones from traps. The recent development of the laminated plastic strip (Hercon[R]) and the hollow fiber (Conrel[R]) dispenser systems has greatly facilitiated research in the use of pheromones. Both of these systems are used by researchers and commercial concerns to dispense pheromones from traps, because they can be engineered to give the desired release rate and longevity under almost any kind of environmental condition.

The technology for formulating pheromones for use in air permeation trials is much less advanced than it is for trapping situations. Many different commercial formulations including microcapsules (NCR[R] and Penwalt[R]), laminated plastic strips (Hercon), and hollow fibers (Conrel) have been tested in the United States and foreign countries during the past several years with varying degrees of success.

Gaston et al. (1977) and Roger Kitterman were able to demonstrate economic control of the pink bollworm in irrigated cotton in desert areas of California and Arizona, respectively, by permeating the air with gossyplure formulated in Conrel fibers. In California, the fibers were fashioned into hoops and applied by hand to ensure that they remained on the plant throughout the growing season. In Arizona, chopped fibers were applied with a machine specifically designed to stick the fibers onto the foliage.

SUMMARY

Scientists working with insect sex pheromones are generally optimistic about their role in pest management schemes. Their most significant use probably will be in the areas of population sampling and monitoring and in the development of techniques for predicting population trends and infestation levels. Application of pheromones and antipheromones for mating control also show considerable promise in many cropping systems. Such studies eventually will lead to improvement in insect pest management with a concomitant reduction in environmental pollution.

REFERENCES

Beevor, P.S., D.R. Hall, R. Laster, R.G. Poppi, J.S. Read, and B.F. Nesbitt. 1975. Sex pheromones of the armyworm *Spodoptera exempta* (Walk.) *Experientia 31:22.*

Brown, E.S., E. Betts, and R.C. Rainey. 1969. Seasonal changes in distribution of the African armyworm, *Spodoptera exempta* (Walk.) (Lep. Noctuidae) with special reference to Eastern Africa. *Bull. Entomol. Res. 58:661-728.*

Byrne, K.J., A.A. Swigar, R.M. Silverstein, J.H. Borden, and E. Stokkink. 1974. Sulcatol: Population aggregation pheromone in the scolytid beetle, *Gnathotrichus sulcatus*. *J. Insect Physiol.* 20:1895-1900.

Campion, D.G., B.W. Bettany, J.B. McGinnigle, and L.R. Taylor. 1977. The distribution and migration of *Spodoptera littoralis* (Boisduval) (Lepidoptera: Noctuidae), in relation to meteorology on Cyprus, interpreted from maps of pheromone trap samples. *Bull. Ent. Res.* 67:501-22.

Croft, B.A. Pest management systems for phytophagous mites and the codling moth. 1975. In Proceedings U.S.-U.S.S.R. Symposium: The Integrated Control of the Arthropod, Disease and Weed Pests of Cotton, Grain Sorghum and Deciduous Fruit. Sept, 28 - Oct. 1, 1975, Lubbock, Texas.

Gaston, L.K., R.S. Kaae, H.H. Shorey, and D. Sellers. 1977. Control of the pink bollworm by disruption of adult moth sex pheromone communication. *Science 196:904-05.*

Marks, R.J. 1976. The influence of behavior modifying chemicals on mating success of the red bollworm *Diparopsis castanea* Hmps. (Lepidoptera, Noctuidae) in Malawi. *Bull. Entomol. Res. 66:279-300.*

McLean, J.A., and J.H. Borden. 1975. *Gnathotrichus sulcatus* attack and breeding in freshly sawn lumber. *J. Econ. Entomol. 68:605-06.*

McLean, J.A.,and J.H. Borden. Suppression of *Gnathotrichus sulcatus* (LeConte) with sulcatol-baited traps in a commercial sawmill and notes on the occurrence of *G. retusus* (LeConte) and *Trypodendron lineatum* (Olivier). *Can. J. Fore. Res. (In press).*

Meijer, G.M., F.J. Ritter, C.J. Persoons, A.K. Minks, and S. Voerman. 1972. Sex pheromones of summer fruit tortrix moth, *Adoxophyes orana*: Two synergistic isomers. *Science 175:1469-70.*

Milligan, R.H. 1970. Overseas wood and bark-boring insects intercepted at New Zealand ports. N.S. Fore. Serv. Tech. Pap. No. 47.

Minks, A.K., and D.J. deJong, 1975. Determination of spraying dates for *Adoxophyes orana* by sex pheromone traps and temperature recordings. *J. Econ. Entomol. 68(5):729-32.*

Minks, A.K., and S. Voerman. 1973. Sex pheromones of the summer fruit tortrix, *Adoxophyes orana*: Trapping performances in the field. *Entomol. Exp. Appl. 16:541-49.*

Minks, A.K., S. Voerman, and J.A. Klun. 1976. Disruption of pheromone communcation with micro-encapsulated antipheromones against *Adoxophyes orana*. *Entomol. Exp. Appl. 20:163-69.*

Mitchell, E.R. 1975. Disruption of pheromonal communication among coexistent pest insects with multichemical formulations. *Bioscience 25:493-99.*

Mitchell, E.R., A.H. Baumhover, and M. Jacobsen. 1976. Reduction in mating potential of male *Heliothis* spp. and *Spodoptera frugiperda* in field plots treated with disruptants. *Environ. Entomol. 5:484-86.*

Mitchell, E.R., W.W. Copeland, A.N. Sparks, and A.A. Sekul. 1974. Fall armyworm: Disruption of pheromone communication with synthetic acetates. *Environ. Entomol. 3:778-80*

Mitchell, E.R., and R.E. Doolittle. 1976. Sex Pheromones of *Spodoptera exigua, S. eridania*, and *S. frugiperda*: Bioassay for field activity. *J. Econ. Entomol. 69:324-26.*

Mitchell, E.R., M. Jacobsen, and A.H. Baumhover. 1975. *Heliothis* spp.: Disruption of pheromonal communication with (Z)-9-tetradecen-1-ol formate. *Environ. Entomol. 4:577-79.*

Moffitt, H.R. 1975. Alternate methods for control of the codling moth. In Proceedings, U.S.-U.S.S.R. Symposium: The Integrated Control of the Arthropod, Disease, and Weed Pests of Cotton, Grain Sorghum, and Deciduous Fruit. Sept. 28-Oct. 1 , 1975, Luccock, Texas.

Riedl, H., B.A. Croft, and A.J. Howitt. 1976. Forecasting codling moth phenology based on pheromone trap catches and physiological-time models. *Can. Entomol. 108(5):449-60.*

Roelofs, W.L., Ada S. Hill, R.T. Cardé, and Thomas C. Baker. 1974. Two sex pheromone components of the tobacco budworm moth, *Heliothis virescens. Life Sciences 14:1555-62.*

Shorey, H.H., L.K. Gaston, and R.S. Kaae. 1976. Air-permeation with gossyplure for control of the pink bollworm. In Pest Management With Insect Sex Attractants, Morton Beroza (ed.). Ann. Chem. Soc. Symposium Series 23. 192 pp.

Shorey, H.H., R.S. Kaae, and L.K. Gaston. 1974. Sex pheromones of Lepidoptera. Development of a pheromonal control of *Pectinophora gossypiella* in cotton. *J. Econ. Entomol. 67:347-50.*

Sower, L.L., and G.P.Whitmer. 1977. Population growth and mating success of Indian meal moths and almond moths in the presence of synthetic sex pheromone. *Environ. Entomol. 6(1):17-20.*

BREEDING INSECT RESISTANCE IN PLANTS:
A CASE STUDY OF WHEAT AND HESSIAN FLY

R. L. Gallun

Department of Entomology,
Purdue University

EARLY CONTROL TACTICS

Breeding resistance to insects in crop plants has made tremendous strides during the past 20 years. The late Dr. Painter and his co-workers showed that insect resistance as well as disease resistance are important components of yield when bred into a crop. Today almost all food and fiber crops have a number of major insect pests attacking them, and breeding for resistance to these insects in crop plants is an integral part of pest management. Indeed, the use of resistant varieties is often the sole method of control. In the case of the Hessian fly, *Mayetiola destructor* (Say), on wheat, the planting of resistant varieties is the main control method used although resistant varieties are sometimes integrated with cultural, chemical, biological and genetic control programs.

The Hessian fly is believed to have immigrated to the U.S. in the wheat straw bed rolls of Hessian soldiers in 1777, hence its name (Havens 1792). Since its introduction this insect has spread throughout the wheat growing area of the United States (Gallun 1964). The adults are mosquito-like in size but their presence in an area can lead to serious losses in wheat. The female oviposits on the leaves of newly planted fall winter wheat in the Midwest and the Great Plains, and the larvae (1/2 mm in size) that hatch from the eggs migrate down the leaf, between the leaf sheaths, to the base of the plant where they begin to feed on the cell sap. During the first four days of feeding the larvae secrete a substance into the plant or take something out that results in the stunting and eventual death of the plant. The larvae continue to feed and grow until they reach a length of approximately 3 mm., then their skins turn a dark color and resemble flax seeds. This "flax seed stage" is the overwintering stage of the Hessian fly in the Midwest and Great Plains states where temperatures remain below 38°F for long periods.

The larvae pupate in the spring and emerge as adults to reinfest the wheat planted the previous fall. Eggs are again laid by the females but this time higher up on the plant since the plant is now in the jointing stage. The larvae migrate down the leaf to the node and feed. There they complete their life cycle by oversummering on the wheat stubble and emerge again in the fall as adults. Thus damage by the second brood of Hessian fly consists of weakening the stem at the sites where the larvae feed. This results in lodging and breaking of straw. Also the sizes of the wheat heads are reduced as are kernal size and weight.

Because of the damage this insect caused in many of the wheat growing states, the U.S. Department of Agriculture and state agricultural experiment stations started studies on its life history and on ways to control it. Different control measures emerged: cultural practices such as proper tillage to turn under the stubble; reduction in volunteer wheat upon which the flies could survive; natural parasitism; safe seeding dates; insecticides; and finally resistant varieties.

It is known that biological control by native parasites will help to reduce Hessian fly populations, but as far as I know a parasite release program has not been initiated. Chemical control with systemic insecticides, both in granular form and seed treatments (Brown 1957), has been worked out and has been effective, but the cost of treating seed, problems of toxicity, and eventual loss of effectiveness has prevented its use.

Safe seeding dates can control Hessian fly populations fairly well. From research it was discovered that wheat could be planted in the fall after most of the Hessian flies emerged from wheat stubble; hence the wheat would escape damage (Davis 1918, Larrimer and Packard 1929). Most of the time this worked, but it did not prevent damage from flies that infested volunteer wheat in the spring. Also there were times when a cool fall kept the flies from emerging until after the fly-free dates. However, safe seeding dates were used quite extensively.

The first resistant wheat variety was developed in the late 1700's (Havens 1792, Fitch 1847). In New York a miller named Underhill planted some of his milling wheat; it proved to be resistant to Hessian fly and demand for his wheat increased. "Underhill" wheat became the popular wheat grown in New York at that time.

It was not until the early 1900's that breeding for resistance to Hessian fly in wheat became an integral part of wheat breeding programs throughout the country. State experiment stations in California, Kansas, Nebraska, and Indiana in cooperation with the U.S. Department of Agriculture, were the forerunners in developing resistant varieties. To date 49 Hessian fly resistant wheat varieties have been developed and released to wheat growers by 11 state experiment stations and two commercial seed companies in cooperation with the U.S. Department of Agriculture.

Hessian fly larvae that feed on susceptible plants cause the stunting of the plant which takes on a dark green color. Larvae that begin to feed on resistant wheat die within a few days, and the plant continues to grow. We still do not know what causes this resistance. It may be a toxin, a nutritional deficiency, or even a non-preferred compound or character that the larvae do not like. In any case, the resistance is genetically controlled and can be transferred to susceptible wheat by plant breeding. Moreover, in wheat we are blessed with many different genes for resistance to Hessian fly and these genes are still being utilized in modern breeding programs. (See Table 1).

MODERN BREEDING PROGRAMS

The breeding of resistant wheats is a team effort that utilizes the skill of different professionals, however, the entomologists and pathologists have similar roles. The entomologist works with resistance to insects whereas the pathologist works with resistance to pathogens. Both work with the geneticist-plant breeder to determine the genetics of resistance and the nature of resistance (Cartwright and Wiebe 1936; Painter et al. 1940; Caldwell et al. 1947; Shands and Cartwright 1953;Allan et al. 1959; Patterson and Gallun 1974).

Laboratory populations of eight biotypes of the Hessian fly are reared in the greenhouse and used for evaluating wheats. We work with seedling resistance although resistance also functions in mature plants. We evaluate wheats, 20 entries to a greenhouse flat, plus checks. Depending on the type of cross made and the generation of the plant populations being evaluated, we score the rows by number of resistant and susceptible plants. This information, and resistant plants, are returned to the breeder for further crosses or selection purposes. During the year we evaluate from 6 to 10 thousand entries for resistance to at least 2 biotypes of fly, sometimes 4. Our Lafayette laboratory cooperates with approximately 5-6 states and three commercial companies. Our Manhattan, Kansas, laboratory has a similar program with wheat breeders in the Great Plains states.

When resistance in a plant has an adverse effect on the life cycle of an insect, then insect biotypes (or races) are likely to develop (Gallun and Reitz 1971). Because resistance in wheat kills the Hessian fly, there is extreme selection pressure in Hessian fly populations for insects that can survive upon the heretofore resistant plant, and these individuals are called biotypes. The biotype situation is analagous to the loss of insecticide control when insects build up resistance to a chemical. When Hessian flies attack wheat having single dominant genes for resistance, there is selection for biotypes having specific genes for virulence or the capability of overcoming resistance. These biotypes are determined by the reaction of wheats having different genes for resistance to the progeny of a single pair mating.

In our work we are using wheats having 4 different genes for resistance as differentials. By doing so we have set up a situation in which 16 biotypes could occur. Of the 16 possible biotypes, we already have nine and are working on more. Table 1 shows the genes for resistance that have been discovered and the plant reactions to biotypes we have isolated or bred.

Gene	Origin	Biotype								
		GP	A	D	C	D	E	F	G	L
H_1H_2	T. aestivum	R	S	S	S	S	S	S	S	S
H_3	T. aestivum	R	R	S	R	S	S	R	S	S
H_5	T. aestivum	R	R	R	R	R	R	R	R	S
H_6	T. durum	R	R	R	S	S	R	S	S	S
H_7H_8	T. aestivum	R	S	S	S	S	S	S	S	S
H_7	T. turgidum	R	R	R	R	R	R	R	R	R

S means susceptable, R means resistant.

Table 1   Reaction of wheats having different genes for resistance to biotypes of the Hessian fly.

One may wonder why we keep developing resistant plants and then develop
virulent biotypes to overcome the resistance. In fact we would be very
happy to avoid biotypes. It certainly would be easy to breed varieties
having one source of resistance and then use this same source of resistance in the breeding programs. As usual, nature does not work this way.
If we do not select biotypes from our laboratory populations and genetic
stocks, natural selection will occur in the field and then we will have
to catch up instead of being able to keep ahead. A good example is the
Indiana release of Knox 62 wheat. Indiana started to use the H_3 gene for
resistance and released a number of varieties having this resistance.
Meanwhile we were using the H_6 gene in other crosses to develop a variety
with a different source of resistance in case the H_3 resistance broke
down to field biotype. Well, in 1962 farmers complained their wheats
were not resistant any more, and we knew Biotype B was becoming prevalent.
Luckily Knox 62 was ready for release on time and Biotype B populations
began to decline as this new wheat was increasingly planted. Unfortunately,
however, Knox 62, though an excellent high yielding wheat with resistance
to Biotype B had a brown chaff the farmers did not like.

Newer varieties, since developed, had better yields and added disease
resistance although they still had the H_3 gene for fly resistance.
Biotype B therefore continued to predominate. Just recently we released
wheats having the H_5 gene that are resistant to all but one biotype. These
wheats are now grown on large acreages in the Midwest and Biotype B is
disappearing. In fact, our 1977 four-state survey of 1322 fields showed

average state infestations of certified wheat fields to be less than 2 percent in each state - a tribute to the growing of resistance varieties. In other words we try to keep ahead by studying the genetic interrelationships that exist between plant and insect by locating new sources of resistance and breeding new biotypes of fly.

The laboratory-bred biotypes are used to seek new sources of resistance to overcome any new biotype formation that appears in the field. They can also be used to distinguish between genes for resistance and to deter - mine whether different genes are combined. Our studies of Hessian fly biotypes have led to a new and exciting method of genetic control. First we have found that the Great Plains biotype has dominant genes for avirulence at every loci comparable to loci carrying genes for resistance in different wheats. We then developed a model that could suppress populations of Hessian fly populations of biotypes in the Eastern United States by the release of the Great Plains fly in the field (Hatchett and Gallun 1967; Gallun and Hatchett 1969). Matings between Great Plains fly and native flies result in progenies that cannot survive on Eastern wheats having genes for resistance. Also progenies of GP x GP matings cannot survive. Only progenies of native x native can survive. By releasing large numbers of GP relative to native fly, suppression to almost eradication can be achieved by a few generations of releases.

When different release ratios of GP to native fly were programmed, the 20 GP:1 native releases almost eradicated the native fly in 3 generations of release in the greenhouse and in the field; the control population remained almost constant and even increased (Foster 1977). This is one more example of integration of resistant varieties with another control method.

SUMMARY

Progress in controlling the Hessian fly has resulted from cooperative programs of breeding for resistance to Hessian fly. Figure 1 shows the results of releasing resistant varieties in Kansas. When there were high acreages of resistant varieties, Hessian fly infestations decreased (determined by number of puparia per 100 culms). However, when this acreage decreased, infestations increased. Figure 2 shows a similar situation in Indiana. Before the release of resistant varieties, field infestations were high. After the release, infestations dropped and remained low.

Table 2 shows the percent of wheat acreages planted to resistant varieties in 1974 in 42 states. During 1974, 14 varieties of resistant wheat were grown on more than 11 million acres in 15 states in the hard red winter wheat region. During the same year 23 varieties of resistant wheat were grown on more than 8 million acres in at least 26 states in the soft wheat region. In the entire wheat region in 1974, 37 varieties of wheats resistant to the Hessian fly were grown in 35 states on more than 20 million acres, or approximately 39 percent of the total wheat acreage grown in the U.S. (Gallun and Briggle from 1974 national wheat survey unpublished).

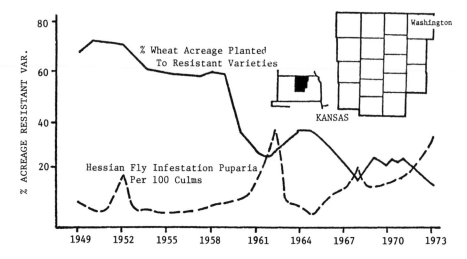

Figure 1 Graph represents a 17 county area of North Central
Kansas infestation based on susceptible varieties.
Percentages of wheat acreage planted to Hessian fly
resistant wheat varieties compared with maximum
Hessian fly infestations 1949 through 1973.
(from Somsen and Oppenlander, 1975)

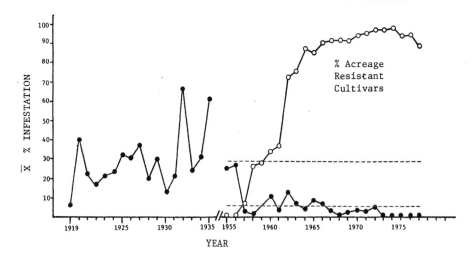

Figure 2 Percentage of wheat acreage planted to Hessian fly
resistant wheat varieties in Indiana compared with
average Hessian fly infestations years 1919-1977.

State	No. of varieties	Hard red winter	Hard red spring	Soft red winter	Soft red spring	White winter	Total
				(Percent)			
AL	11	0.0	0.0	44.6	0.0	0.0	44.6
AZ	0	0.0	0.0	0.0	0.0	0.0	0.0
AR	13	0.0	0.0	90.5	0.0	0.0	90.5
CA	0	0.0	0.0	0.0	0.0	0.0	0.0
CO	5	54.2	0.0	0.0	0.0	0.0	54.2
DE	6	0.0	0.0	81.2	0.0	0.0	81.2
FL	3	0.0	0.0	85.7	0.0	0.0	85.7
GA	4	0.0	0.0	51.1	0.0	0.0	51.1
ID	0	0.0	0.0	0.0	0.0	0.0	0.0
IL	17	15.0	0.0	74.8	0.0	0.0	89.8
IN	11	.1	0.0	98.4	0.0	0.0	98.5
IA	12	57.9	2.4	0.0	0.0	0.0	60.3
KS	11	47.5	0.0	0.1	0.0	0.0	47.6
KY	8	0.0	0.0	83.1	0.0	0.0	83.1
LA	4	0.0	0.0	51.8	0.0	0.0	51.8
MD	5	0.0	0.0	78.7	0.0	0.0	78.7
MI	10	0.0	0.0	23.4	0.0	27.3	50.7
MN	0	0.0	0.0	0.0	0.0	0.0	0.0
MS	4	0.0	0.0	52.9	0.0	0.0	52.9
MO	15	9.9	0.0	74.1	0.0	0.0	84.0
MT	3	3.8	0.0	0.0	0.0	0.0	3.8
NB	11	52.6	0.0	0.0	0.0	0.0	52.6
NV	0	0.0	0.0	0.0	0.0	0.0	0.0
NJ	5	0.0	0.0	69.6	0.0	0.0	69.6
NM	3	37.3	0.0	0.0	0.0	0.0	37.3
NY	2	0.0	0.0	0.4	0.0	0.6	1.0
NC	8	0.0	0.0	75.9	0.0	0.0	75.9
ND	1	0.1	0.0	0.0	0.0	0.0	0.1
OH	14	0.0	0.0	95.6	0.0	0.0	95.6
OK	9	20.0	0.0	0.0	0.0	0.0	20.0
OR	0	0.0	0.0	0.0	0.0	0.0	0.0
PA	9	0.0	0.0	74.8	0.0	0.0	74.8
SC	7	0.0	0.0	31.2	0.0	0.0	31.2
SD	4	10.7	0.0	0.0	0.0	0.0	10.7
TN	11	0.9	0.0	85.6	0.0	0.0	86.5
TX	8	8.5	0.0	1.4	0.0	0.0	9.9
UT	1	1.3	0.0	0.0	0.0	0.0	1.3
VA	8	0.0	0.0	77.6	0.0	0.0	77.6
WA	0	0.0	0.0	0.0	0.0	0.0	0.0
WV	8	0.2	0.0	66.0	0.0	0.6	66.2
WI	9	1.6	1.5	2.8	0.0	0.0	5.9
WY	5	28.8	0.0	0.0	0.0	0.0	28.8
No. states in which grown		18	2	25	0	3	35

Table 2 Percent of State wheat acreages grown to Hessian fly resistant wheats in 1974.

87

Yield studies have shown that wheat growers save at the least one bushel per acre of wheat by growing resistant varieties. If wheat sells in the range of $2 to $3.50 per bushel, we have saved approximately $40-70 million in one year; a pretty good return on the investment in agricultural research.

So this is the status of Hessian fly resistant varieties today. We have many resistant varieties that are being grown in the U.S., and they are doing the job they are supposed to be doing; protecting the crop. We must always keep on top of things and search for new sources of resistance and monitor for new biotypes. Team work is essential in the breeding program. Integrated control methods such as safe seeding dates, good cultural practices, the release of parasites, and genetic control can and should be used in connection with resistant varieties. Many researchers and farmers have made this program a success.

REFERENCES

Allan, R.E., E.G. Heyne, E.T. Jones, and C.O. Johnston. 1959. Genetic analysis of ten sources of Hessian fly resistance, their interrelationships and association with leaf rush reaction in wheat. *Kansas Agric. Exp. Stn. Tech. Bull. 104, 51 pp.*

Brown, H.E. 1957. Hessian fly control with systematic insecticides. *F.A.O. Plant Protect. Vol 5(10):149-155.*

Caldwell, R.W., W.B. Cartwright, and L.E. Compton. 1946. Inheritance of resistance derived from W38 and durum PI 94587. *J. Am. Soc. Agron. 38(5):398-409.*

Cartwright, W.B. and G.A. Wiebe. 1936. Inheritance of resistance to the Hessian fly in wheat crosses Dawson x Poso and Dawson x Big Club. *J. Agric. Res. 52:691-695.*

Davis, J.J. 1918. The control of three important wheat pests in Indiana. *Indiana Agric. Exp. Stn. Circ. 82:1-11.*

Foster, J.E. 1977. Suppression of a field population of the Hessian fly by release of the dominant avirulent Great Plains biotype. *J. Econ. Entomol. 70:775-778.*

Gallun, R.L. 1964. The Hessian Fly. *Rev. U.S.D.A. Farmers Bull. 1627: 1-9.*

Gallun, R.L. and J.H. Hatchett. 1969. Genetic evidence of chromosome elimination in the Hessian fly. *Ann. Entomol. Soc. Am. 62(5):1095-1101.*

Gallun, R.L. and L.P. Reitz. 1971. Wheat cultivars resistant to races of Hessian fly. U.S. Dept. of Agric. ARS Prod. Res. Rep. 134:1-16.

Hatchett, J.H. and R.L. Gallun. 1967. Genetic control of the Hessian fly. Proc. N.C. Branch Entomol. Soc. Am. 22:100 (abstract).

Havens, J.N. 1792. Observations on the Hessian fly. Soc. Agron. New York Trans. Part I: 89-107.

Larrimer, W.H. and C.M. Packard. 1929. Hessian fly control in Indiana. *Indiana Agric. Exp. Stn. Circ. 167:1-12.*

Painter, R.H., E.T. Jones, C.O. Johnston and J.H. Parker. 1940. Transference of Hessian fly resistance and other characteristics of Marquillo spring wheat to winter wheat. *Kans. Agric. Exp. Stn. Tech. Bull. 49:1-55.*

Patterson, F.L. and R.L. Gallun. 1974. Inheritance of resistance of Seneca wheat to Race E. of Hessian fly. Proc. 4th Int. Wheat Genetics Symp. Mo. Agric. Exp. Stn. (August 1973): 445-449.

Shands, R.G. and W.B. Cartwright. 1953. A fifth gene conditioning Hessian fly response in common wheat. *J. Am. Soc. Agron. 45(7):302-307.*

Somsen, H.W. and K.L. Oppenlander. 1975. Hessian fly biotype distribution, resistant wheat varieties and control practices in hard red winter wheat. USDA. ARS-NC-34 (December) 7 pp.

THE ROLE OF CHEMICALS
IN INTEGRATED PEST MANAGEMENT

B. G. Tweedy
CIBA-GEIGY Corporation

INTRODUCTION

It is a real pleasure for me to meet with you and discuss integrated pest management. I have had the opportunity to work for a university, the USDA and now with industry. I have also had the opportunity to visit several countries, including the USSR, where both quality and quantity of food is a great problem. Mr. de Jong has requested that I speak about the role of agricultural chemicals in pest management programs.

I would like to begin by defining what I mean by integrated pest management. I think of IPM as "the use of multiple control measures which are compatible, economical, environmentally sound and culturally feasible for managing pest populations at an acceptable level." In some countries this "acceptable level" is somewhat different than in the U.S.; total production is emphasized with less emphasis on quality of commodity produced. In the U.S. we want both high production and high quality. In the Soviet Union the use of pesticides in pest management practices is increasing. Ours is beginning to level off or decrease. I would like to point out at this time that IPM is a U.S. term and most other countries use IPC or Integrated Pest Control.

To many people, IPM means the use of only biological control measures. To others, IPM implies the use of non-chemical control measures. However, to most of the scientists and growers with whom I have been associated, IPM means the use of several measures for controlling pests with a minimum impact upon the environment. The correct use of chemicals is certainly one of these control measures. This was brought out in the workshops yesterday and today. Included in the definition of the word "correct" is the choice of the right chemical, applied at the right rate and at the right time. When chemicals are used in this manner, I believe we have nothing to be ashamed of. However, we should be concerned when we misuse chemicals. I believe the misuse of chemicals has led to most of our current pestcide problems of public interest. By misuse, I am referring to having inadequate knowledge available for defining the correct use. I am not referring to illegal use because of use outside of the label.

Today we read so much about the adverse effects of chemicals to our environment, but little is said about the benefits of agricultural chemicals.

These benefits are very impressive and I wish I had time to discuss them in detail. Dr. Pimentel made this point quite clear yesterday. The use of agricultural chemicals has been, and is, a major factor in the U.S. being the leading producer of high quality food and fiber with a minimum number of people on the farm. The American farmer produces food for himself and over 70 other people. The "other people" are freed from the farm to work in other industries, or to become doctors, teachers, scientists, or whatever they choose. No other country in the world enjoys the high quality of food at the low cost, or low percent of total salary, as we do in the U.S.

Secretary Bergland was quoted in a written version of his talk to the National Agricultural Chemicals Association (NACA) on September 27, 1977 as stating that "there appears to be an opportunity to help ease this nation's unemployment by using 'people power' instead of only chemicals in our fight against pests." In Case Report 68 it was stated that controlling weeds in corn by hand labor at the average farm labor price of $2.65 per hour in 1976 resulted in a net loss of $66 per acre. The net profit due to use of a herbicide was $78. In 1977 the price of corn went down and the cost of labor went up, thereby making an even greater differential. In addition, there aren't enough unemployed people to control weeds in U.S. crops even if they wanted to hoe weeds on the farm. I feel it is a gross oversimplification to say that the way to solve unemployment is to put the people back on the farm with hoes and fly swatters to control pests.

Another implication was made by Bergland that "in these energy-short times" we are being unrealistic by using petroleum-based products to manufacture pesticides. Information available from the U.S. Department of Agriculture indicates that the increase in productivity from the use of pesticides exceeds the total of agricultural exports. Also, this same source indicates that there is a 70-fold return in energy used for the petroleum-based pesticides.

COST/BENEFITS OF CHEMICALS

I believe the benefits of agricultural chemicals have outweighed and continue to outweigh the costs. I do not deny that there have been costs. One must also remember that costs are apparent with other signs of "progress" such as new highways, the emissions from cars and airplanes, suburban developments and many other insults which we have placed on our environment. The key to the use of agricultural chemicals is to use them so they have the maximum beneficial effect and the least negative effect on our environment. Using them in sound pest management programs certainly has merit and is being more actively researched now than in past years.

Emphasis by industry, universities and government research groups is being placed upon the more correct use of pesticides. As an industry, we now know much more about what happens to our pesticides when they are applied to the environment before we apply for registration. For example,

92

we know how rapidly and how far chemicals leach in different soils, how rapidly they break down in the environment, how toxic or safe they are to fish, wildlife, mammals, and many other non-target organisms. We are in a much better position than ever to define how to use chemicals judiciously. We know that by using smaller amounts of two or three chemicals applied in combination, we can frequently get better pest control. While the total chemical applied per acre may frequently be more, the effect on non-target organisms and the levels of residues is usually much less. Some of the above research problems were at one time considered basic research done only by a university.

Where are the real problems with chemical control? In the past, chemical control in many instances has been used to attempt total elimination of the pest from a crop. Nature has provided built-in systems to by-pass such road blocks, so it should not be surprising to us that pests have adapted to these measures in various ways and to varying degrees. Pest resistance, however, is nothing new with the advent of organic chemicals. Nature has provided us with pests which attack previously resistant varieties of crops for many, many years. An example is the breeding for resistance of wheat to wheat rust. *Puccina Graminis*, the causal organism of wheat rust, has for several decades been outdoing the plant breeder and a resistant wheat variety has remained resistant for only three to four years.

Today we recognize that total control or absolute elimination of pests is not essential. Today scientists and farmers recognize that the real goal is to keep the pest population below an "economic threshold". We recognize we can put up with some level of the pest, but we must control it to a degree that it does not destroy our crops, or ruin the crop to the point that it is uneconomical to the grower. Yes, growers as well as the chemical industry must make a profit in order to survive. Thus, the correct use of fertilizers, agricultural chemicals, selection of crop varieties, biological control, crop rotations and other cultural methods, breeding programs, irrigation practices, etc., all fit into this program. No one factor, including chemicals, can do it all. It must be a program using all the tools and resources available.

Controlling pests is a challenging exercise in applied ecology. In order to achieve our goal of economical crop production and sound environmental practices, we must first have a good understanding of the agroecosystem in which the crop is grown. This is essential to defining a successful pest management program. We must know which pests occur every year and cause an unacceptable amount of crop injury and which pests are secondary because they generally do not cause significant yield losses. Pesticides play a key role in controlling each of these types of pests and I believe we are in a better position now than ever before to more correctly define the use of pesticides in most of these pest control programs without causing any adverse effects. With a better understanding of the various agroecosystems, we are approaching a new horizon for agricultural chemicals in pest management systems. We now have a better understanding of what types of compounds are desirable, and needed, in integrated pest management programs. The unfortunate thing that has occurred is that the cost of developing agricultural chemicals is so great that industry can no longer afford to develop them for some of these special uses.

INDUSTRY VIEW ON IPM

How does the chemical industry see integrated pest management programs? Here again we must refer to the definition of IPM. Pest management research programs, supported for many years by federal and state funds, were accelerated in the early 1970's. Their aim was to reduce the pesticide load in the agroecosystem. Although the programs were called pest management programs, they were essentially insect management programs designed to scout crops such as cotton and apples in order to make better decisions on when insecticides should be applied. Where properly carried out, they have been quite successful in both pest control and the economics of crop production. As you saw from Dr. Frisbie's talk this morning, better cotton varieties have been integrated into the system and the cotton pest management program has also been improved.

The National Agricultural Chemicals Association (NACA), whose membership is composed of companies that synthesize and manufacture most of the agricultural chemicals used for pest control in the United States, has had a keen interest in the pest management program development. In 1972, the following policy was adopted by the NACA membership:

NACA endorses and urges support of programs which have as their ultimate objective the achievement of pest suppression based on sound ecological principles which integrate chemicals, biological and cultural methods into a practical program, where necessary and when possible.

One of the main objections the agricultural chemical industry has toward the pest management programs that were developing was, industry was not invited to participate in the planning of such programs or allowed to give inputs where our proprietary compounds were concerned. We felt no one knew more about the performance and safety of the chemicals used in pest control than the companies that developed them. We could have made, and can now make, very valuable inputs into such programs. However, the chemical industry is looked upon as an avid opponent of pest management because the companies would "lose revenue." This is far from reality because most pest management programs do include the use of agricultural chemicals and such programs will bring about the need for newer and better chemicals. This need presents an opportunity for the agrichemical industry. We are interested in our environment and have been active in the national pesticide monitoring program since 1967.

Many stresses work to overcome man. Yet, the nearly four billion people on the earth today have about 20% more food per person than did the 2.7 billion people of 20 years ago. Are we satisfied with this record? No. Many millions of people are still starving.

94

America is a land of beauty - especially where food is grown. And
our farmers have been good stewards of the soil. But we do have
problems. Every field has problems, often hidden stress that farmers
must overcome. All stress begins with the soil, sunshine, and moisture,
because all life depends on them. God has slowly, but surely allowed
us to create tools to cope with stress. We should not throw them away,
but we should learn to use them more wisely, and build on them. Good
seed, fine agricultural machines, fertilizer, agricultural chemicals,
better cultural practices, and biological control are all tools we cannot
neglect if we expect to survive. This is what modern agriculture is
all about -- survival. We cannot go back to the 1800's or even the early
1900's. We must continue on a course of progress. We have made a lot
of mistakes, but we've done a lot of things right. We have a good agricultural industry, let's improve it. We have a high quality environment,
let's improve it. The two objectives are compatible and, for the
benefit of our children and their children, they are essential.

DISCUSSION

QUESTION: *What is the involvement of the pesticide indsutry in monitoring
the fate of chemicals in the environment?*

TWEEDY: *CIBA-GEIGY has been monitoring the presence of our chemicals
in water and in soil·for quite some time and the industry as a whole has
also been.*

QUESTION: *We have been talking about integrated pest management. I spent
many years developing a very successful integrated control program on
cotton in California. The program was initiated after a disaster occurred
with cotton bollworm which was precipitated by chemical controls used against
the lygus bug. Then, two or three years ago, there was an advertisement
for an insecticide called Supracide. It was broadcast on a Central Valley
radio station and it went something like this: "Mr. Farmer, the lygus
bug is about to invade your fields. At the very sight of the lygus bug
start spraying with Supracide. Get on a regular program and you will
have a cleaner crop and more profit." How do you equate this kind of
thing with the glorious views and interests expressed by the agrichemical
industry about IPM?*

TWEEDY: *We are trying to be much more careful about how we recommend
the use of pesticides in pest management programs. As a matter of fact,
we just recently took one off the market that was very good because we
were not sure how to use it. This was a voluntary action on our part. Since
that time we have done a considerable amount of research and found it does
fit into a pest management program. We put it back on a limited basis for
a particular pest management project. I do think we're becoming more
responsible as an industry, but, I won't deny that we are going to continue to try and make a profit by selling a product and will continue to
encourage farmers to buy that product.*

PART FOUR
IMPLEMENTATION

IMPLEMENTATION OF INTEGRATED PEST MANAGEMENT PROGRAMS

Leon Moore

Extension Entomologist,
University of Arizona

INTRODUCTION

What is pest management? What makes it suddenly so important--even popular? How and when did it start? Why should I understand it? Is it compatible and usable in today's technical agriculture?

Pest management brings together into a workable combination the best parts of all control methods that apply to a given situation. A somewhat more scientific definition of pest management would be: the practical manipulation of pest populations using sound ecological principles. The emphasis here is on *practical* and *ecological*. There are many ways of controlling insect pests, only a few of which are practical, and fewer yet ecologically sound; that do not create a worse situation. Pest management then, is "putting it all together" --using the best combination of control techniques that permits us to "live" with the pest while sustaining non-economic losses.

Pest management as a concept is not new; only the name is. Many of the components of a sound pest management system were known some 50 years ago through the research of Isely in Arkansas. His extensive and foresighted work with cotton insects in the mid 1920's was sufficient to provide a sound basis for today's pest management. His management of such pests as the boll weevil, the bollworm, and spider mites, was based on the principles of *applied ecology*, a vital component of pest management.

Why is insect pest management needed? Why not continue to control insects as they have been in the past? The answers to these questions are somewhat complicated and yet they must be dealt with and understood.

Introduction of the organochlorine insecticide, DDT, began an era of insecticidal control of insects. Entomological research and extension work largely emphasized the use of insecticides to control insects. One new insecticide followed another and new groups such as organophosphates and carbamates made their appearance. Insecticidal control provided a quick, inexpensive and convenient method of controlling insects. It greatly slowed or stopped efforts such as Isely's to develop methods of insect control which were forerunners to the methods which we are using in our insect pest management systems today.

99

There were many reasons for the need to resume emphasis on the development of insect pest management. These were brought to light as problems began to occur resulting from large scale use of insecticides.

One of the first problems was the development of resistance or tolerance by certain insects to insecticides used against them. Beginning with the resistance of houseflies to DDT, this problem has continued to increase until today about 250 species have shown resistance to certain insecticides and some are resistant to one or more groups.

After a few years of widescale use of insecticides, the problem of residues remaining in food and feed crops, in the soil, and in animals became known. Some insecticides such as the organochlorines are highly persistent because of their chemical stability. Others such as the organophosphates are less persistent and rapidly degrade into harmless compounds. In order to cope with the residue problem, growers found it necessary to use the more toxic but non-persistent compounds.

The shift from persistent to non-persistent insecticides has helped to relieve the problem caused by remaining residues but has been largely responsible for the occurrence of other problems. The non-persistent insecticides are generally more toxic, creating an additional health hazard to persons handling and applying them. They also are generally broad spectrum, i.e. toxic to many insect species in the treatment area, and require more frequent applications to maintain insect control. This has resulted in a disturbance of pest-beneficial insect relationships, permitting pests of minor importance to rise to major pest status. It has also resulted in increased costs since the less persistent insecticides are generally higher priced and more applications are required. These factors have contributed to the need for developing pest management systems which emphasize alternative methods of control and minimize the use of insecticides.

BASIC ELEMENTS OF INSECT PEST MANAGEMENT

Four elements basic to the development of a pest management program are sampling, economic levels, natural control, and insect biology and ecology. A good sampling system is extremely important in that it provides information on insect numbers in each field and must be developed to serve as a base for utilizing knowledge of natural control, economic levels, and the biology and ecology of the major insects involved. Once the sampling program is established, these basic elements can be dovetailed together to serve as the foundation upon which practical components can be added to the total pest management program.

Before a pest management program can be initiated a great deal of basic information must be accumulated. This includes information about the agroecosystem, such as the crops grown, agronomic practices employed, soil type, irrigation water, and any other factor which relates to the production of the crops in the system. Detailed information must be available on the major pests and beneficial insects found in the agroecosystem in order to understand the seasonal occurrence and magnitude of all species of concern.

The integration of all information on the agroecosystem itself with
that on the biology and ecology of the pests and beneficial insects will
provide significant insight on natural control in any particular area.
The level of natural control provides the base on which all management
practices are built, some of which enhance natural control.

PRACTICAL COMPONENTS OF INSECT PEST MANAGEMENT

When the basic elements have been established to form the foundation
for the insect pest management system it is possible to build a solid
and effective program on this base. There are several single-component
control methods that can be incorporated into a multifaceted insect pest
management system. These methods have been, for the most part, used
individually for control of specific insect pest problems. The combination of several of these into a comprehensive insect pest management
program can provide better suppression of key pest species, and, at the
same time, place less demand on any one method. The methods currently
available and proved effective are: cultural control, biological control,
chemical control, host-plant resistance, mechanical-physical control, and
regulatory control.

The number of components that can be used in an insect pest management
system is limited only by their practical availability. If the available
components are to be used most effectively, emphasis must be placed on
their use at the appropriate time. Some components are applied when the
pest is a problem in the field while others are applied at times when the
pest is overwintering or when it is at sub-economic levels. Generally,
full utilization of all non-chemical methods should be emphasized on a
year-round basis and insecticides should be utilized as a means of reducing
populations that have reached or exceeded the economic level.

Several potential components are in various stages of development at the
present time. These include pheromone control, microbial control, chemo-
sterilant control, and other control methods. These should become important parts of insect pest management systems as they are developed to the
point of being practical for use.

EXAMPLE OF AN INSECT PEST MANAGEMENT PROGRAM: COTTON IN ARIZONA

Emphasis in cotton pest control programs in Arizona has been aimed primarily at developing or adapting a cotton scouting program to the state
which would serve as the basis for an insect pest management system. Two
prerequisites to the effective practice of pest management are: good field
sampling; and confidence in and use of sound economic levels of pest populations or damage.

Four pest management programs involving four major Arizona cotton producing
counties were conducted in 1977. Counties involved included Graham, Pinal,
Pima, and Maricopa. About 50,000 acres of cotton were included in the
programs which were entirely grower financed. In addition to cotton, the
program in Pinal county continued the multicrop approach and involved
7,400 acres of other crops including small grains, sugar beets, alfalfa,

and grain sorghum. The Pinal County Growers Pest Management Corporation employs a full-time supervisor and part-time secretary and bookkeeper in addition to the necessary scouts to keep their program on a year-round basis. A new program established under a private consultant in Maricopa county operates on a year-round basis and involves other crops grown in the area. Programs in other areas operate only during the cotton growing season. Growers continue to request Extension Service involvement in the operation and development of their programs.

Grower involvement has increased each year since the pilot program was initiated in Pinal county in 1971. About 95% of the 1977 growers followed pest management principles compared to 22% when the program started in 1971. Some of the changes brought about include:

1. Increased involvement of growers in pest management;
2. Establishment of grower organizations to operate pest management programs;
3. Full financing of programs by growers;
4. Increased number of private consultant firms for pest management purposes;
5. Year-round practice of pest management through the multicrop approach;
6. Utilization of the sex pheromone gossyplure for early season mass-trapping and monitoring of pink bollworm males;
7. Better utilization of naturally occurring beneficial insect populations in pest management;
8. Use of resistant varieties;
9. Harvest management of alfalfa to reduce lygus migration into cotton;
10. Treatment of safflower with insecticides to reduce lygus numbers and the subsequent problem in cotton.

Acceptance of pest management by growers is indicated by its continued growth in Arizona. The new program established in Maricopa county gives pest management good exposure in all the major crop producing areas of the state. About 150 growers participated in 1977 compared to 140 in 1976 and cotton acreage increased from 44,000 to 50,000 acres. Other crop acreage amounted to approximately 7,400 acres.

All programs are completely grower administered and financed although the new programs still require much Extension assistance. Two grower groups are operating as cooperatives while two programs are operated by contracting with private consultants. The growers are eager to promote the development of pest management as evidenced by their acceptance and financing of the sex pheromone trapping addition to the programs in 1975, 1976, and 1977. This addition will be continued in 1978 at grower expense.

Grower benefits from IPM increase as the program in a county matures. This is probably due to growers becoming more familiar with pest management objectives and to greater confidence in program personnel. Growers become more involved each year they participate and this also adds to grower benefits. Of special interest is the overall reduction of grower costs by reducing insecticide use which also contributes to environmental quality. In Pinal county on the average number of treatments per acre on 20, 761 program acres in 1977 was about 7.8 compared to approximately 9 in 1971 when the program began. 1977 was a severe cotton insect year and 15 or more treatments were common by growers outside the programs. The 44 program growers in Pinal county spent an average of $54.87 per acre for insect control while it is estimated growers outside the program in Maricopa, Pinal and Yuma counties spent an average of $100.00 per acre. In comparing treatment costs one must keep in mind that insect populations and the need for control varies from year to year. Pest management enables the growers to take advantage of light infestation years, however, as well as as reducing pest control costs during years of serious pest outbreaks.

In Graham county, only 6,500 of 15,000 acres were treated in 1977. All the acreage was treated on a pre-scheduled basis prior to establishment of the pest management program in 1969. Growers are pleased with the program because of the low cost for insect control and the improved relationship they have with the general public in their communities. Another spin-off from the program has been improvement in the honeybee industry. One beekeeper stated that the program increased his income about $20,000 a year. Tables 1 and 2 give a summary of IPM Program results in Graham and Pinal counties respectively.

	1973	1974	1975	1976	1977
Program Acres	5,487	11,076	7,634	8,014	15,560
Sprayed Acres (Total)	45,618	4,930	1,957	290	25,510
Scouting Cost	$ 7,806	$18,386	$13,823	$16,685	$32,000
Spraying Cost (Material and application)	$155,100	$26,946	$11,032	$ 1,538	$127,590
Pheromone Trap Cost			$15,267	$32,056	$70,020
Total Cost	$162,906	$45,332	$40,122	$50,179	$229,610
Total Cost Per Program Acre	$29.69	$4.09	$5.26	$6.26	$14.75

Table 1. Information Summary for Graham County IPM Program

	1971	1972	1973	1974	1975	1976	1977
Growers	51	60	85	54	35	41	44
Fields	387	480	722	503	295	435	471
Acres	15,260	19,313	31,582	21,458	12,742	19,172	20,761
Fields Not Treated	---	12	29	16	8	57	0
Acres Not Treated	---	95	982	687	165	2,022	0
Acres Treated	6,390	18,115	29,995	20,431	12,577	17,150	17,845
Total Acre Treatments	56,232	173,330	156,563	106,252	75,222	65,772	139,890
Range of Treatments	1-13	0-16	0-12	0-10	0-11	0-10	1-15
Avg. Number Treatments Per Acre	8.8	8.9	5.2	5.2	5.7	3.4	7.8
Total Cost of All Acre Treatments	$196,812	$606,655	$626,252	$531,260	$451,322	$394,632	$979,230
Avg. Total Cost Per Acre Treated	$ 30.80	$ 33.49	$ 20.88	$ 26.00	$ 35.88	$ 23.01	$ 54.87
Est. Cost of Material And Application/Acre	$ 3.50	$ 3.50	$ 4.00	$ 5.00	$ 6.00	$ 6.00	$ 7.00

Table 2. Information Summary for Pinal County Pest Management Program

ECONOMICS OF PEST MANAGEMENT

Raymond Frisbie

Department of Entomology,
Texas A&M University

It is indeed a pleasure for me to attend this Pest Control Strategies
Conference to discuss the economic implications of pest management
programs with you. I have been requested to address my remarks primarily
to the area of cotton integrated pest management programs. This area of
evaluation is most comfortable to me and I think has, as we will see in
subsequent discussion, been most productive in the area of economic evaluation of both research and implementation of IPM strategies. There will
be no attempt to define integrated pest management. This concept has
been defined many times and should be more properly presented as a
philosophy rather than a clear cut definition. The responsibility of all
IPM systems is to present the participants involved with a fair and a
reasonable profit. In addition, IPM has a responsibility to man and his
environment, i.e., to minimize as much as possible the introduction of
hazardous chemical toxicants into the environment. With these two broad
goals in mind, we should address the area of economics with the basic
assumption that most of the strategies developed for IPM are done so within the economic concept.

The understanding of the economic threshold is basic to IPM systems. If
insect populations are allowed to increase to economically damaging levels,
then there is usually no other alternative than the application of an
insecticide to prevent or reduce economic loss. IPM strategies must bring
all population suppression measures to bear prior to the time populations
develop and pass the economic threshold level. The components of an IPM
system should be integrated in such a way that the various pest population
suppression or regulation factors would ideally prevent or minimize the
use of insecticides. However, the state of sophistication of most IPM
systems, particularly in cotton pest management programs have not reached
the stage of sophistication where chemical insecticides are not required.
In certain instances, IPM systems have been developed that use less insecticides with relatively little environmental disturbance. However,
in those areas where populations have risen above economic threshold levels,
chemical insecticides have and will for some time in the foreseeable
future play a key role in the management of insect populations.

COTTON IPM PROGRAMS

The economic and environmental assessment of IPM programs must be included
as a key component in all research and implementation stages. A careful
economic track record must be kept to determine the feasibility of the

various research and implementation approaches. Cost/benefit ratios
should be developed and a clear indication of the impact of IPM strategies
on net profit must be determined. Several excellent examples of the
economic feasibility of IPM research systems have been demonstrated. This
research was developed with a systems approach to better understand
the various components involved in the economical management of insect
pests. An IPM system in the Lower Rio Grande Valley of Texas utilized
a short-season, indeterminate cotton variety (Tamcot SP-37) in combination with reduced fertilizer and irrigation use, and field scouting
reports to assist in decisions to apply insecticides on an as-need-basis
(Namken and Heilman 1973). Additional research also included a variation
in the row width utilizing the short-season cotton genotype. This study
was compared with conventional methods of irrigated cotton production in
the Lower Rio Grande Valley. Insecticide costs decreased by $10.14
per acre under the IPM program. Annual insecticide costs for conventional
cotton were approximately $28.91 per acre compared to $18.77 for the
short-season, narrow-row system. Net returns for conventional production
were estimated to be $37.27 per acre while under the IPM program net
returns were estimated to be $55.77, or an increase of $18.50 per acre
(Lacewell et al. 1977). A similar short-season, narrow-row cotton study
was conducted in Frio County Texas in 1974. An evaluation of this study
using the short-season cotton Tamcot SP-37, narrowing the row width to
26 inches, and reducing nitrogen fertilizer levels and irrigation water,
showed that pesticide applications were reduced to an average of 6.6
applications as compared to conventionally grown cotton that averaged
16.9 (Sprott et al. 1976). Further analysis indicated that the short-
season, narrow-row system returned a net profit of $252 an acre as com-
pared to $109 per acre for conventional cotton grown on 40 inch rows.

A research and demonstration project conducted in the Trans-Pecos of
West Texas has provided some interesting opportunites for economic production of cotton (Lindsey et al. 1976, Condra et al. 1978). An economical production system was developed for the Pecos River Valley. A
complete economic study was conducted prior to the initiation of the
plan juxtaposed with the production factors that impinged on the growth
and management of cotton. This system was termed ECONOCOT and was
developed out of a need to increase profitability of cotton production
in the Pecos River Valley. Increased prices in natural gas for well
pumps, high insecticide use and overall inflated production costs have
produced a gradual decline in the cotton acreage in the Pecos River
Valley since the early 1970s. A research-demonstration study that
included all variables in the economics of cotton production was considered and the results of that were quite revealing. Short-season
indeterminate cotton varieties (Tamcot SP-21), an intermediate
maturing cotton variety (McNair 612) and 2 long-season indeterminate
varieties (Stoneville 213 and Deltapine 16) were compared under different
management schemes. Using reduced fertilizer and water inputs along
with timed insecticide applications based on economic threshold information, the total pest management package demonstrated quite clearly
that the short-season cotton Tamcot SP 21 returned $364.38 per acre
compared with net returns generated for Stoneville 213 and Deltapine
16 of $134.49 and $108.19, respectively (Lindsey et al. 1976).

Perhaps one of the best examples of the economic impact of an applied pest management program is seen in the Texas Cotton Pest Management Program conducted through the Texas Agricultural Extension Service. Texas Cotton Pest Management Program objectives included the judicious use of pesticides based on economic thresholds as determined by field inspection (scouting), conservation of beneficial insects, introduction of new technology into a total systems approach for the overall management of insect pests. The results of the economic evaluation for the Texas Cotton Pest Management Program have been well documented (Frisbie et al. 1975). In a comparison of farmers participating in the Texas Cotton Pest Management Program with a similar group of non-participating farmers, total net profits for 1973 and 1974 in the 35,000 acre program increased approximately $2,100,000 due to increases in yield and decreases in insecticide use, or both. Concurrently there was a reduction of 82,000 lbs. of pesticides entering the environment during the 1973 and 1974 production season (Frisbie et al. 1974). A similar economic evaluation was used to evaluate the expanded Texas Pest Management Program in 1976. Increased net returns were calculated to be $5,594,000 for 100,000 acres of cotton included in the Texas statewide IPM program in 1976 (Frisbie 1978 unpublished data).

SUMMARY

The results of these evaluations clearly indicate that IPM has a strong economic and environmental base. As we proceed forward in the implementation of IPM programs, it is essential that the position of IPM be clearly placed in the context of a total agricultural production cropping system. It is also strongly suggested, as we develop into the area of medical/veterinary and urban entomology, that similar evaluation schemes be developed prior to or at least during the early stages of implementation.

DISCUSSION

QUESTION: *How do you get farmers to make the transition to short-season cotton?*

FRISBIE: *Our philosophy is that it is best to deal with an educated man. We spend alot of time and energy educating the farmer. If you can show the farmer a way to make more money and farm better, he will most likely listen. The cost of insecticides, particularly if you get into a bollworm fight, are just prohibitive. We have areas of Texas where they chose to go bankrupt rather than change and we have areas of Texas where they are changing radically to this short-season system. Yields have increased and cotton production across large areas has stabilized. Insecticide use has gone down and the industries in those areas have come back into viability.*

REFERENCES

Namken, L.N., and N.D. Heilman. 1973. Determinate cotton cultivars for more efficient cotton production on medium textured soils in the Lower Rio Grande Valley of Texas. *Agronomy Journal Vol. 65 pp. 953-956.*

Lacewell, R.D., J.E. Casey, and R. Frisbie. 1977. An evaluation of integrated cotton pest management programs in Texas: 1964-1974. Departmental Technical Report No. 77-44 Texas Agricultural Experiment Station. pp. 34-44.

Sprott, J.N., R.D. Lacewell, G.A. Niles, J.K. Walker, and J.R. Ganaway. 1976. Agronomic, economic, energy and environmental implications of short-season, narrow row cotton production. Texas Agricultural Experiment Station, MP-1250. 24 p.

Lindsey, K.E., G.D. Condra, C.W. Neeb, L. New, H. Buehring, D.G. Foster, J. Menzies. 1976. Texas ECONOCOT system upland cotton demonstration in Pecos County 1976. Texas Agric. Ext. Serv. Unpublished memo.

Condra, G.D., K.E. Lindsey, C.W. Neeb, and J.L. Philley. 1978. ECONOCOT...A Ray of Hope for the Pecos Valley. *Texas Agricultural Progress. Vol 24:3p.*

Frisbie, R.E., J.N. Sprott, R.D. Lacewell, R.D. Parker, W.E. Buxkemper, W.E. Bagley, and J.W. Norman, Jr. 1975. A practical method of economically evaluating an operational cotton pest management program in Texas. *J. Econ. Entomol. Vol 69:2 pp. 211-214.*

Frisbie, R.E., R.D. Parker, D.E. Buxkemper, W.E. Bagley, and J.W. Norman, Jr. 1974. Texas Pest Management Annual Report, 1974, Cotton, Texas Agricultural Extension Service. Unpublished mimeo.

EMERGING FEDERAL POLICIES ON PESTICIDES

Charles Reese

Office of Pesticide Programs,
Environmental Protection Agency

HISTORICAL PERSPECTIVE

Recently chemicals were developed which gave the American farmer a
means of controlling pests at low cost. Some of these chemicals
provided spectacular results and were persistent enough to give long-
term crop protection, causing many users to drop the more traditional
preventative forms of pest control. This increased dependence on the
use of pesticides has led to pest resistance, secondary pest problems,
undesireable crop residues, and non-target effects. Federal policies
developed since World War II resulted in pesticides being the major
control tool available for use by pest managers. These successful
policies were related to cheap and abundant supplies of land and energy.
Today, less land, increased energy costs and environmental concern
necessitate a shift in Federal policies. A look at past pesticide
policies and programs is necessary to understand the emerging Federal
policies concerning pesticides.

In 1910 the Federal Insecticide Act gave the Federal government the
authority to remove fraudulent or misleading materials from the market.
In 1947, the Federal Insecticide Fungicide and Rodenticide Act was
passed to regulate the marketing of economic poisons and devices. Together with amendments made in 1959, 61 and 64, the Federal Insecticide
Fungicide and Rodenticide Act defined the term "economic poison" as
having the same meaning as the more commmonly used term "pesticide."
It is defined in the Act as "any substance or mixture of substances
intended for preventing, destroying, repelling or mitigating any insects,
rodents, fungi, weeds and other forms of plant and animal life viruses,
except viruses on or in living man or other animals" declared to be
a pest by the Administrator and "any substances or mixture of substances
intended for use as a plant regulator, defoliant, or dessicant."

"Devices" are mechanisms such as ant traps, sold together with pesticides
for the purpose of application; or simply mechanisms such as electronic
bug-killers, designed to destroy pests.

Under the Federal Insecticide Fungicide and Rodenticide Act, the United
States Department of Agriculture required that:
1. All pesticides shipped interstate be registered. Adulterated,
   misbranded or insufficiently labeled products were prohibited
   from interstate commerce.

2. Registration was granted for five years when test data proved the pesticide safe and effective when used as directed on the proposed label.
3. Use directed on the label for food and feed could not result in pesticide residues greater than proposed residue tolerances-until 1970 under the Food, Drug and Cosmetic Act - Health, Education and Welfare.
4. Labels of highly toxic chemicals were required to contain the word poison and describe the antidote.

In 1954, the Miller Amendment (Sect 408) to the Food, Drug and Cosmetic Act authorized the Food and Drug Administration of the Department of Health, Education and Welfare to set tolerances on pesticide residues on raw food. Some pesticides were registered on a negligible residue basis. As the techniques of chemical analysis became more sensitive, residues were detected and it became necessary to decide if these newly discovered residues were a hazard to public health. In 1965 the National Academy of Sciences - National Research Council recommended that the concept of negligible residues as used in the registration and regulation of pesticides be abandoned. A joint United States Department of Agricuture-Health, Education and Welfare implementation of the National Research Council report was published in the Federal Register on April 13, 1966. It was agreed that registrations of all uses involving reasonable expectation in the absence of a finite tolerance or exemption would be discontinued as of December 31, 1967. Many registrants did not submit for tolerance for certain crops; as a result, many uses were cancelled. Other registrations for zero tolerance pesticdies continued on the basis of pending petitions for a finite tolerance or on the basis of progress reports on ongoing studies. A 1974 report found that the resolution of this question was a social decision.

In 1958, the Delaney Clause was added to Sect 409 of the Food, Drug and Cosmetic Act. This addition to the food additives section states that no food additives capable of causing cancer when ingested by animals or man may be added to food. Then in 1969, the National Environmental Policy Act was passed. The National Environmental Policy Act requires Federal agencies which use pesticides to incorporate a concern for the quality of the environment into agency missions. In addition, the National Environmental Policy Act established the Council on Environmental Quality. The Council on Environmental Quality supervises environmental impact statements which require Federal agencies to consider alternative actions, solicit advice from other Federal agencies with expertise and consult with Council on Environmental Quality. These environmental impact statements have had an effect on Federal pesticide use policies. Specifically, the application of pesticides in water areas or non-problem areas is now being avoided.

Reorganization Plan #3 established the Environmental Protection Agency December 2, 1970. The Environmental Protection Agency was designated as the central Federal pollution abatement agency responsible for the protection of the environment against all types of harmful pollution,

specifically including pesticides. The Environmental Protection Agency
has been given pesticide regulatory responsibilities previously scattered
through a number of Federal agencies. The transfer of responsibilities
included:
1. Pesticide Registration from the United States Department of
   Agriculture.
2. Tolerance setting for pesticide residues on food and feed from
   the Food and Drug Administration of the Department of Health,
   Education and Welfare.
   a. Responsibility for enforcement of pesticide residues on raw
      agricultural products remains with the Food and Drug Ad-
      ministration while like responsibilities for pesticide
      residues in meat and poultry in interstate and foreign
      commerce rests with the United States Department of Agri-
      culture (Animal Plant Health Inspection Service).
   b. Regulation of pesticide product advertising is the re-
      sponsibility of the Federal Trade Commission.
   c. The Department of Transportation is responsible for pesticide
      packaging.
3. Certain technical assistance and research functions from the
   Public Health Service of the Department of Health, Education and
   Welfare.
4. From the Department of Interior:
   a. Federal Water Pollution Control Act functions
   b. Pesticide Research Act functions
   c. Activities of the Gulf Breeze Biological Lab.

As you know, the Environmental Protection Agency at its inception focused
on the hazards of pesticide pollution. At a time when an increasing
Federal role was called for, the Environmental Protection Agency was operating
under enabling legislation designed at an earlier time for programs of
other agencies with somewhat different missions.

This situation set the stage for the amendments to the Federal Insecticide,
Fungicide and Rodenticide Act in 1972 and 1975 which gives the Environmental
Protection Agency authority to:
1. Control all pesticide use.
2. Classify pesticides.
3. Approve certification of applicators of restricted use pesticides-
   Department of Transportation (Federal Aviation Administration)
   is responsible for aerial applicators.
4. Conduct research on biologically compatible alternatives for
   pest control.
5. Make integrated pest management information available.

These congressional mandates make clear the responsibility Federal agencies
have to use and advise the use of pesticides in the most efficient,
environmentally sound manner possible. In his environmental message to
Congress, the President has asked for the development of a Federal policy
on integrated pest management. The Council on Environmental Quality at
the conclusion of its ongoing review of integrated pest management in the
United States has been asked to recommend actions which the Federal government could take to encourage the development and application of pest

management techniques which rely on chemical agents only as needed. In this regard, Secretary Bergland has pointed out that the need for a systematic approach to commodity protection based on sound economic, ecological, technical and societal considerations is essential for maintaining agricultural production in the United States. He went on to say that "Each system must be economical and compatible with other farm, forest and urban horticultural management practices." I would like to briefly summarize on-going Federal agency activities in the development of responsive pesticide policies.

FEDERAL AGENCY ACTIVITIES IN IPM

Council on Environmental Quality (CEQ)

As an advising body to the President, the Council on Environmental Quality provided the first Federal Policy statement on integrated pest management. The Council on Environmental Quality has been asked to update the status of integrated pest management and then recommend to the President policies and actions the Federal government could take to encourage the development of integrated pest management.

United States Department of Agriculture/University Complex

Historically, the vast majority of pest control research, development and implementation has originated from this source. Programs conducted include development of techniques of biological, cultural, chemical, and crop varietal methods of control as well as pest population surveillance. Recent programs have been initiated aimed at gaining farmer acceptance and use of integrated pest management. The United States Department of Agriculture has the lead responsibility for the development of new integrated pest management technologies.

Department of Health, Education and Welfare (HEW)

Health, Education and Welfare along with several universities here and in Canada has been working to improve career opportunities in integrated pest management through symposia, curricula development and vocational training. Comprehensive integrated pest management curricula for community and junior colleges have been developed. Health, Education and Welfare in cooperation with other Federal agencies is developing an integrated pest management education program for secondary schools. The program includes basic information dissemination and general education of urban and rural dwellers concerning their ecosystem. Pest problems and their control are to be part of full scale implementation of a comprehensive integrated pest management educational program. This program here is an example of Health, Education and Welfare initiatives.

Department of Housing and Urban Development (HUD)

This department has become increasingly involved in the further development and implementation of integrated pest management. Along with developing a manual on pest management, Housing and Urban Development

112

is planning for the development and implementation of integrated pest management in public housing and especially as part of housing rehabilitation projects.

National Science Foundation (NSF)

The National Science Foundation has been a continual supporter in the development and implementation of integrated pest management programs. In cooperation with the Environmental Protection Agency and the United States Department of Agriculture, the National Science Foundation has sponsored projects to test the practical use of Integrated Pest Management.

Department of State, United States Agency for International Development

The United States Agency for International Development (AID) is collaborating with the University of California (UC) to develop and utilize integrated pest management methods for crop protection throughout the world. Its objectives are to promote ecologically sound crop protection tactics, to develop grower capabilities for making sound pest management decisions, to improve the socioeconomic position of farmers by increasing the quantity and quality of commodities produced to the consumer and to maintain the quality of human life.

National Academy of Sciences (NAS)

The National Academy of Sciences has had a strong interest in the development of pest control technology. Over the past few years the Academy has played a major role in promoting integrated pest management programs by making available several publications outlining the latest pest management methods. The National Academy of Sciences continues to provide valuable assistance through its continual assessment of the state of development and implementation of pest management programs.

Department of Defense (DOD)

Working through the Armed Forces Pest Control, the Board Department of Defense develops pest control manuals, trains pest control personnel and is responsible for pest control related to military bases and activites.

Environmental Protection Agency (EPA)

The Environmental Protection Agency has been carrying on programs related to the development and implementation of integrated pest management programs. In the developmental phase, this Agency has the responsibility to: (1) Register and Control the use of pesticides and (2) develop biologically integrated alternatives for pest control. As a part of implementation, Environmental Protection Agency continues to make instructional materials concerning integrated pest management available upon request in cooperation with the United States Department of Agriculture, the Council on Environmental Quality, the Department of Housing, Education and Welfare and the Department of Health, Education and Welfare.

Integrated Pest Management may prove to be a way of avoiding the cancellation of some pesticide uses in the areas of experimental use permits and later the registration process itself will be examined as a means of responding to public concern and Agency responsibilities.

Some Federal pesticide use policies which may be expanded are:
1. One Federal policy on integrated pest management - not an agency by agency effort.
2. Organization of pest control research and development as pest control systems.
3. Improved access to all information pertaining to pest, pest control techniques and integrated pest management systems. While integrated pest management is not an across the board panacea, existing management options should be available to those concerned and certainly problems should be known to those with responsibilites or incentives to solve or reduce them.
4. Expansion of integrated pest management educational and training programs.
5. An announced tilt toward integrated pest management by United States Department of Agriculture.
6. Provide profit oriented growers with timely, useful information and incentives.
7. A better definition of registration requirements for pheromones and hormones.
8. Private sector involvement.

SUMMARY

There are interactions between pest management decisions, the environment and society. Management decisions effect the environment, commodity production and production costs. While the pesticide user is usually profit motivated, society also weighs environmental effects. Society therefore, has been prompted to establish regulations and promote the availability of management information for pesticide users.

Policy developed since World War II resulted in pesticides being the major control tool available for use by pest managers. This successful policy was related to cheap and abundant supplies of land and energy. Today, environmental concern, less land, and increased energy costs necessitate a shift in policy. It has been demonstrated that integrated pest management systems can operate under these new constraints while maintaining the urban and rural environment and increasing agricultural productivity. The President has called for a national integrated pest management strategy. The Environmental Protection Agency will be working closely with the Council of Environmental Quality, the United States Department of Agriculture, the Department of State(AID) and other concerned agencies in a positive continuing response to the President's request. This emerging Federal policy concerning pesticide use is responsive to public concern, legislative mandates and developing technology.

DISCUSSION

QUESTION: *You mentioned the requirements regarding the use of pheromones and hormones. What are those requirements?*

REESE: *Some people feel these third generation pesticides are "good" things and should be zipped through the registration process. If, on the other hand, you have to sit in the registration division, you probably have a different feeling about it, a different kind of responsibility. Desmon Johnson, who heads the pesticide program, is pushing very hard to get this and other parts of the registration process moving ahead faster. We have been held up by court cases and disagreement over different points of view. Whatever the compound, we want to see that people are aware of what the requirements are and also that these materials move through the registration process faster.*

USDA PERSPECTIVES ON PEST MANAGEMENT

R. L. Ridgway

Science and Education Administration,
United States Department of Agriculture

MISSION

I appreciate this opportunity to share with you some thoughts on pest management from the perspective of the U.S. Department of Agriculture. As you know, USDA is involved in a number of pest control activities including: (1) mission-oriented fundamental research, (2) applied research to develop pest control tactics, (3) research to develop decision-making technology to aid in determing when control methods are needed, (4) integrating tactics into management systems, (5) evaluation of economic feasibility, (6) extension and technology transfer, and (7) regulatory and action programs. These activities are conducted in cooperation with other Federal and State agencies and the private sector.

HISTORY

Pest management and integrated pest management may be new terms to many, but some of the underlying concepts have long been recognized. Before the advent of modern synthetic organic pesticides, cultural and biological methods of controlling pests were common, and they remain today the primary means of controlling disease and nematodes, and, to a lesser extent, insects and weeds. For example:
  (1) A lady beetle predator that completely controlled the cottony cushion scale on citrus in California was imported by USDA in 1888.
  (2) Varieties of wheat that were rust resistant were developed in the early 1900's. Later, resistant varieties of wheat were integrated with optimum planting dates to control Hessian fly.
  (3) In the 1930's an integrated approach was used in a cooperative Federal-State-grower program to control the phony peach disease.

However, the synthetic organic pesticides that were available following World War II provided effective, economical, and convenient pest control. As a result, pesticide use increased rapidly and played a dramatic role in increasing agricultural production. Less emphasis was therefore placed on nonchemical pest control. But even during this period, pest control on forage and small grain crops continued to emphasize cultural and varietal resistance because these methods had such low cost.

Indeed, as early as the late 1950's, the USDA expressed concern over problems that were arising because of pest resistance to pesticides and the adverse effects of pesticides on the environment. Therefore, at this time, the Federal Research unit (formerly the Agricultural Research Service) of the Science and Education Administration began to revise its research program. By 1970 approximately 80 percent of its insect and plant disease control budget was directed toward fundamental biology and alternative methods of pest control. However, the general effectiveness and low direct cost of pesticides (herbicides, insecticides, fungicides, nematicides) compared with alternative control methods, encouraged the expansion of pesticide use until, at the present time, over one billion pounds per year are being applied in the United States. Market value of pesticides has also greatly increased.

The public concern over pesticide use in this country continues to increase and has resulted in the development of comprehensive pesticide regulations to protect environmental quality and human health. Also, pest resistance to pesticides has increased substantially. The net result is the great interest in pest management, which may reduce pesticide use, pest control costs, and health risks and also may result in improved environmental quality. In addition, it may delay the obsolescence of individual pesticides due to the development of resistance. More emphasis is now being placed on selecting, integrating, and using pest control tactics based on anticipated economic, environmental, and sociological consequences rather than on routine pesticide treatments.

USDA ROLE IN IPM

The U.S. Department of Agriculture strongly endorses the concept of integrated pest management. The Department will continue to develop, practice and encourage the use of those tactics, systems, and strategies of practical, effective, and energy-conserving pest management that will result in protection against the environment. Thus, the Department will stress integrated approaches to pest management problems in its research, extension, regulatory, and action programs. In the process, the Department will be mindful of the interests and pest management needs of all segments of American society including those interested in gardens, households, small farms, commercial farms, forests, food and fiber handling, and storage and marketing enterprises.

In order to insure top level policy pest management support, Secretary of Agriculture Bob Bergland issued a Secretary's Memorandum on pest managment. He emphasized "We will be placing increased emphasis on controlling significant pest populations with biological and other natural controls as well as with selective chemical pesticides." However, he added, "the policy should not be interpreted as a move to eliminate the use of the pesticides that U.S. agriculture is dependent upon, because they are part of the integrated pest management approach...The policy statement should be seen as an increased concern by the U.S. Department of Agriculture for the health and well being of all Americans and for the ecosystem of which we are a part."

As part of the Department's efforts to seek more desirable approaches to pest control, USDA sponsored a special study team to review the status and prospects of biological agents for pest control. This joint Federal-State-industry effort provided a useful basis for future activities. The final report contains 11 recommendations aimed at expanding the use of biological agents. Perhaps the following 2 examples will provide some insight into actions needed to increase the use of biological agents:

(1) Expand the USDA's competitive grants/contracts research program in order to focus expertise on existing or emerging solutions to problems through the use of biological control agents. Such a program would provide support for qualified scientists, wherever located, including educational institutions, research foundations, private investigators, commercial enterprises, and Federal and State agencies.

(2) Provide additional technical assistance to potential users of biological control agents and assess the need for various types of incentives to private enterprise to encourage and hasten participation in the development and use of biological agents for pest control. Such assistance should be made available to pest management consultants, grower cooperatives, and commercial production and distribution enterprises.

Although these and other needs were identified to encourage the development of biological controls, many of these same principles apply to other desirable control tactics such as pheromones and insect growth regulators.

Clearly, the USDA is committed to integrated pest management, and we are aware of many changes that need to be made for more effective implementation. However, for integrated pest management to be effective, it must become an integral part of a large number of different agricultural production systems. The increasing diversity of agricultural systems adds considerable complexity to the development of integrated pest management systems for all situations. For example, traditional commercial farms, now numbering about 2.7 million, are larger, more mechanized, and fewer in number than ever before. At the same time, the number of farms concentrating on alternative agricultural production methods has increased to about 500,000 and the number of home gardens now exceeds 35 million. The more labor-intensive agriculture production systems such as home gardens offer some unique opportunities for integrated pest management. More emphasis in this area will require the involvement of a larger number of people.

As we look toward future integrated pest management in all segments of agriculture and forestry, we see a need for strengthening research and development, technology transfer, economic assessment, and implementation. Substantial resources are currently available for these activities. It is important that we work with all interested parties in order to effectively utilize these resources. The Executive Budget for fiscal year 1979 recommends modest increases in integrated pest management funds for biological controls, host plant resistance, and pest management for small farms. With modest increases in resources and more effective use of current resources, we are looking forward to the expanded use of integrated pest management.

DISCUSSION

QUESTION: *Why do you place more emphasis on integrated pest management in home gardens than you do for ornamental plants?*

RIDGWAY: *At the present time, there is probably more support in USDA for expansion in the home gardening aspect since it is more closely related to food production. Also, there is a new major thrust on small farms proposed within the USDA. In the Executive Budget for fiscal year 1979, there is a 3.5 million dollar increase request focusing on small farm technology. Much of this effort can impact on home gardeners. In addition, the Extension Service is currently funding a number of home garden pilot projects. I have been putting some of my personal efforts into involving USDA in the small farm and home garden area. I feel this is the easiest place to make the transition into urban pest management. In preliminary planning for the increased research, we have identified a number of linkages that can be made with interested groups that USDA has not actively served in the past. However, at the present time, a number of our land-grant universities, including Cornell University and Penn State University, are conducting research in the home gardening area.*

It is my hope that USDA can presently do more about pest management on horticultural crops and in home gardens. If Congress supports an increase in this effort beginning in 1979, an increased effort will follow on pest management for ornamental plants.

MAKING THE TRANSITION
TO AN URBAN IPM PROGRAM

Helga Olkowski and William Olkowski

*Center for the Integration of the Applied Sciences,
John Muir Institute*

INTRODUCTION

Since 1971 the authors have been developing programs to manage the
plant-pest-human interactions in urban areas. We began with urban
shade tree insect pest problems. Cooperating with us were the Parks
and Recreation and Public Works Departments of five cities in the
north central coastal and Sacramento valley of California (1,2).
During the last two years the project expanded to examine indoor and
structural pest problems, working with a school district (3). Currently
the Center for the Integration of Applied Sciences (CIAS) is involved
in studying the management of a range of insect and disease problems
on non-tree ornamental vegetation as well as vegetables in backyard
and community gardens (4,5).

Our concern is the overuse and misuse of pesticides in urban areas.
Since integrated pest management (IPM) programs in agriculture generally
demonstrate pesticide use reduction, it seemed logical to explore the
application of the IPM approach to these other settings. At this point
we conclude that it is entirely possible and desirable to develop urban
IPM programs and that the consequent reduction in the use of toxic
materials is likely to be substantial wherever this is done (5). However,
the characteristics of urban areas require the modification of agricultural
models for such programs to be successful.

THE URBAN CONDITION

To briefly summarize the relevant differences between the two human-
designed systems, urban and agricultural: urban areas are characterized
by a greater density of people; greater diversity of vegetation and
microclimates; overlapping pesticide use patterns and jurisdictions
(for example, the same piece of turf may be treated by the homeowner,
the mosquito control agency and drift from sprays aimed at the municipal
shade trees); and, to a very great degree, the pest problems in urban
areas are those of nuisance or aesthetics rather than of economic con-
sequence. Where pest problems resemble agricultural situations the most,
as in backyard and community food fardens, the small-scale and recreational

121

nature of the systems makes feasible and desirable intensive care and alternative strategies to pesticide use.

The mineral and plant resources of the nation come from the countryside, pass briefly through the hands of the city-dweller, and then make their way to the dump, often the nearest body of water. Pesticides are an example of a resource that follows this route. However, they differ from other toxic materials that are inadvertently released into the environment as a by-product of manufacturing processes or urban life-styles. Most pesticides are compounds that have been deliberately con-structed to interrupt and destroy living systems. Thus their use in areas of great human density, particularly by relatively untrained entomophobic homeowners, janitors, gardeners and the myriad of others either casually or extensively involved in urban pest control, is a special category of resource management.

THE COMPONENTS OF AN URBAN IPM PROGRAM

From the above description of the salient aspects of urban areas it can be seen that educational efforts are a necessary feature of IPM implementation. Furthermore, research in urban IPM technology transfer, from the IPM specialist to the political and maintenance personnel of the system to be managed, must take on a distinctly interdisciplinary approach. The ecologist-IPM specialist finds that incorporation of techniques of analyses and integration, from such varied discipines as sociology, psychology, political science, public education and business management, becomes a necessary requirement (7,8).

The programs developed by our project so far have all had three major components: delivery system, education and research. The delivery system includes the monitoring of potential pest insect populations, their natural enemies, and human behaviors that affect the pest problem. The latter includes other horticultural activities such as watering, fertilizing, pruning or mowing, mulching, plant selection, etc., human food storage and waste management methods, and systems for training, deploying and communicating with personnel directly or peripherally in-volved in pest management. The monitoring system provides the informa-tion necessary to set up a communication and training system for imple-mentation of intervention strategies to suppress pest populations when and where necessary.

The research component involves determining what levels of the various pest populations require treatments ("injury levels"), development of alternàtive strategies suitable for use against the various pest pro-blems that arise, and evaluation of treatments so that a predictive capacity is developed within the system. Every effort is made to deter-mine the best methods of enhancing the natural biological controls that are already present in the system. Where the biology of the pest and/or its natural enemies is inadequately known it may be studied through the monitoring process already in place through the delivery system as well as by means of lab cultures and field experiments.

Education is the key component of the system. The people maintaining the vegetation or habitats must be trained to recognize the natural enemies of the pests, make counts upon which timing and site selection of treatments can occur and incorporate alternative pest management strategies into their ongoing programs. The general public must also be educated since it is frequently their aesthetic value judgements or ignorance that triggers pest management actions.

SEQUENCE FOR ESTABLISHING AN IPM PROGRAM

During the first year in a new system the monitoring process is initiated to determine which are actual pest problems and which are triggered by previous pest or other horticultural management techniques. A history of treatments and an inside picture of the bureaucracy involved is obtained. Problems are rated as to their severity, and a priority list for focusing efforts is agreed upon by all involved. Usually specific areas of low visibility, where pest damage can be tolerated, are set aside to help in determing injury levels and presence of biological control agents. The educational program is initiated through a system of regular reports to personnel involved in immediate and supervisory management activities.

Perhaps most important, the geographic, biological and bureaucratic boundary of the IPM program is decided upon during this first season. It is essential that this boundary be set to encompass a large enough area to permit the solution of the problems included in the system. What is being done in one area may affect problems in another. For example, turf management may affect the surrounding trees, the way one species of shade tree is treated may affect pest management upon other tree species in the city, students handling of snacks and organic wastes in a classroom may affect the cockroach problem in the area, etc.

We began working with the city of Berkeley on a classical biological control project against the linden aphid (*Eucallipterus tiliae*) under the auspices of the Division of Biological Control, University of California, Berkeley. The city had requested control of that problem specifically and we proceeded, in the usual manner of university researchers attempting biological control in an orchard or alfalfa field, to focus solely on the importation of a specific parasite (*Trioxys curvicaudus*). After nearly losing study sites through pesticide treatments of adjacent and different vegetation we began to perceive that the city-maintained trees were a system both bureaucractically and biologically. Predators of insects moved from tree to tree just as city tree maintenance crews do. By drawing the boundaries of the project large enough to encompass the management of all the trees in the city we were able to successfully colonize and spread the natural enemy of the aphid with consequent permanent solution to that particular problem. In addition, the satisfactory development of management strategies for a whole series of other problems was thereby also obtained. The result was the eventual substantial reductions in pesticide use throughout the city.

Because of the vagaries of weather, and variation in other maintenance practices, it usually takes at least two seasons and sometimes a third (for example, when two dry years are followed by a wet one as recently happened in California) to establish injury levels with any certainty. During the second and third years, alternative strategies that have proven effective in sample areas are adapted to larger portions of the system that include significant variations of microclimate, soils, habitat or human use, etc.

When the various major pest problems of significance in the system are examined, one or more may be found to be caused by an insect invader that has left its natural enemies behind in its area of origin. If no practical alternative strategies to pesticide treatments can be found to suppress the populations of the invaded insect the feasibility of importing its natural enemies should be examined (10). The authors have employed this technique successfully using host-specific parasites of various aphids. This strategy offers the potential of an important partial or total solution to particular problems (12). Another biological control approach may involve the use of insect diseases, such as *Bacillus thuringiensis* (Biotrol^R, Dipel^R, Thuricide^R) which is specific against certain caterpillars and does not disturb the beneficial insects.

It should be stressed, however, that some natural biological control occurs all the time, even in an area heavily treated with pesticides or in a well-kept indoor environment (for example, we have found parasites on cockroaches inside school buildings). The reason that humans are able to survive on this planet and grow any plants at all is because most insect species are under good natural biological control by predators, parasites and disease. Insect populations that cause problems are those that for one reason or another are inadequately suppressed by the natural controls under present conditions. The statement that IPM means "the integration of cultural, biological and chemical control methods" has thus confused some people into assuming the biological control component must mean starting a classical importation project. In fact, the primary efforts of an IPM program are directed towards the preservation and enhancement of whatever biological controls may already be operating in the system. This is why a careful monitoring process is so important.

During subsequent seasons major efforts are made to transfer the technology to the maintenance people in the system. A predictive capacity should be in place to aid in increasing efficiencies of labor and materials use. With reasonable certainty regarding the reliability of various management strategies, a vigorous public education effort also can be made. To someone encountering the idea of an IPM program for the first time, the complexity of the decision-making process, skills required to adapt new methods or integrate them into former approaches, and the length of time needed to establish an ongoing program, may make undertaking the effort seem prohibitive. In fact, it appears to us that in most urban systems it will take a trained IPM consultant several seasons to establish a workable program. Furthermore, the first program in a bio-region may need support for applied research beyond the financial means of the cities involved. In addition to direct support from the cities and a school district, our pioneering work is also currently funded by the Environmental Protection Agency and the State Department of Water Resources and Food and Agriculture. However, it is our experience that once the program has been developed the technology can easily be transferred over to the city or institution for which it was developed. The process for doing so can be designed into the system from the start.

In the final analysis, the adoption of such urban IPM programs will depend on several factors: the public's increasing awareness of the long-term health and environmental hazards from pesticide use, the perception by various institutions of the increasing costs and restrictions on the employment of these materials, and the realization on the part of the policy makers that viable options already exist (13).

DISCUSSION

QUESTION: Are birds considered a part of the natural predator complex?

ANSWER: There are, of course, many insectivorous birds in urban areas. I think the important thing to understand is that as you reduce pesticide use you allow more of the natural controls like birds and smaller predators to survive. Any program that is able to reduce pesticide use allows for more natural controls to operate. You are using the ones that are already there. You don't have to pay them, they work on weekends and when you are back in the office. To use all the natural controls available is one of the major aims of a good integrated pest management program.

QUESTION: As you are evaluating the parasites and predators that are to be imported from foreign countries are you also evaluating their potential interactions with the environment?

ANSWER: Absolutely. Let me provide an example: there are about 6,000 species of aphids in the world. Taxonomists break it into about 13 tribes. When we look at natural enemies of these aphids, there are not only natural enemies that attack only aphids but some that are specific to certain genera of aphids. There are natural enemies that will attack only one species and not its close relative. I don't want the polyphagous species (that attack a broad range of hosts) because they are opportunists - they go where the food is. I want the specific parasites because they will best regulate a pest population. Specific parasites will not get involved with other components of the system. It is extremely complicated, like a metabolic fit between two compounds. It is not comparable to introducing starlings or even ladybird beetles. Also there are strict quarantine procedures that must be followed.

QUESTION: How economical is your shade tree program?

ANSWER: *The city of Berkeley estimated that we saved them $22,500
in 1972. City governments don't fire anybody, they just relocate labor.
Most of the cost of treating in cities involves labor; pesticides are
a small cost. The savings were arrived at by taking treatment costs
per tree and multiplying it by the number of trees that they didn't
have to treat after our program started. If you combine the total
number of trees from our other city programs in California the total
savings approached $200,000. Our program saves them a small amount
of money even when consultants are included. What is not economical
for them is to pay for the cost of the initial research and development
of an IPM program. That is why we get EPA and state help. We set up a
model program in a particular region, then your private consultant can
pick up that information and start using it in other cities that have
similar kinds of problems.*

*QUESTION: How do you see your urban program affecting the agricultural
community?*

ANSWER: *We see our efforts in urban areas as a flanking movement
to affect policy makers and the general public. 73.5% of the people
in the U.S. live in urban areas according to Bureau of Census information. We have some indications that this information has gone over
to the agricultural sector. This is one of our tactics and I can see
it starting to work.*

REFERENCES

Note: The following references are cited for the purpose of providing
the reader with additional information regarding points mentioned in
this paper.

1. Olkowski, W., D. Pinnock, W. Toney, G. Mosher, W. Neasbitt, R. van
    den Bosch, and H. Olkowski. 1974. A Model Integrated Control
    Program for Street Trees. *Calif. Agr. 28(1):3-4.*
2. Olkowski, W. et al. 1978. Urban Integrated Pest Management, In:
    Pest Control Strategies. E. Smith, D. Pimentel (Eds.). Academic
    Press.
3. Olkowski, H., W. Olkowski, K. Davis, L. Laub. 1978. Developing an
    Integrated Pest Management Program for a School District. Proceedings
    of the XII Annual Conference of the Association of Applied Insect
    Ecologists, Newport Beach, California.
4. Olkowski, W. and H. Olkowski. 1975. The City People's Book of
    Raising Food. Rodale Press, Emmaus, Penn. 228 p.
5. Olkowski, H., W. Olkowski. 1975. The Integral Urban House,
    to be published Autumn , 1978.
6. Olkowski, W. and H. Olkowski. 1977. Developing Urban IPM Delivery
    Systems. Paper delivered at IPM Conference: New Frontiers in Pest
    Management, Sacramento, California. Proceedings to be published.
7. Olkowski, H. and W. Olkowski. Sept. 1976. Entomophobia in the
    Urban Ecosystem. *Bull. Entomol. Soc. Amer. 22(3):313-317.*

8. Olkowski, W., H. Olkowski, R. van den Bosch and R. Hom. 1976.
    Ecosystem Management: A Framework for Urban Pest Management.
    Bioscience 26(6):384-389.
9. Olkowski, W. 1973. A Model Ecosystem Management Program. Proc.
    Tall Timbers Conf. Ecol. Anim. Control Habitat Manage. 5:103-117.
10. Olkowski, W., H. Olkowski, A. Kaplan, R. van den Bosch. 1978.
    The Potential for Biological Control in Urban Areas: Shade Tree
    Insect Pests, In: Perspectives in Urban Entomology. J.W. Frankie
    and C.S. Koehler (Eds.). Academic Press.
11. Olkowski, W., and H. Olkowski. 1976. Integrated Pest Management
    for City Trees. Proceedings of the Midwestern Chapter of the
    International Society of Arboriculture. Pp. 21-31.
12. U.S. Department of Agriculture. 1978. Biological Agents for
    Pest Control: Status and Prospects. U.S. Dept. of Agriculture
    and the Agricultural Research Institute. 138 pp.
13. Olkowski, H. and W. Olkowski. 1978. Some Advantages of Urban
    Pest Management Programs and Barriers to Their Adoption.
    Proc. of the IPM Seminar presented by the University of California
    Cooperative Extension Service in cooperation with the L.A. Commissioner's
    Office and the Southern California Turfgrass Council.

CURRENT AND FUTURE RESEARCH NEEDS

Kenneth Hood

Office of Research & Development,
Environmental Protection Agency

INTRODUCTION

I would like to compliment Pieter de Jong and the Wright-Ingraham
Institute for the work which has gone into this well organized conference. The program has moved along nicely and I think we all owe a debt
of thanks to Pieter and the stalwart crew which has assisted him.

I am in the Office of Research and Development at the Environmental
Protection Agency and I was asked to talk on the current and future
IPM research needs. With a topic which looks into the future, my greatest
need is for a fortune teller's clear crystal ball. I could not find
one to borrow. Nevertheless, I shall give you some thoughts I have
been gathering over the last several days on what I think we are going
to eventually need in IPM research.

Integrated Pest Management is comparatively new on the government scene,
as far as EPA is concerned. We believe there are both rural and urban
IPM needs. For the immediate future in agriculture, I do not believe
the direction of present research is going to significantly change from
what we are now pursuing. Our major emphasis now is on insect control
but there is beginning to be a gradual awareness of a need to investigate
plant or weed control, which I will discuss a little later.

INSECT CONTROL

Regarding insect control research, I believe that in the future,
it will be necessary to know much more about many of our important insect
pests. When we examine what is now being done in any crop system, we
often find most of the work is concentrated on an intensive study of
just a few of the problem insects. I believe we will see an expansion
of study into more species.

There will be an expansion of work into insect population dynamics
of our worst pest species and their interactions with others. It will
be necessary to have more biological information on how insects live and
the effects of the environment upon them because they are very adaptive

to adversity. The gene pool with which we are dealing has sufficient resiliency that it has survived thousands of years. Mankind's pest management efforts represent just another stress for it. We are going to have to do more research on the crop-pest interactions. Oftentimes the entomologists or the agronomists follow their own narrow speciality and do not talk together and compare notes. We will find increasing need to bring the disciplines closer together.

WEED CONTROL

Plant control is an old subject that is again being noticed. We have heard discussions about no-till and reduced-till culture, the use of herbicides, and so forth. We are still heavily dependent on chemical control. There is a lot being done in cultural weed control but if energy supplies become a problem, tractors cannot be run as often as needed and the farm community is going to ask "What are our options? What else can we do?" This situation will bring to the forefront the need for non-chemical weed control which might include utilizing pathogens and predators. These are comparatively unexplored fields and there are not many researchers working full time in these areas.

We need a lot of information on field ecology dealing with interactions between useful and non-useful plants; that is, crops and weeds. Why are weeds so competitive? There are many reasons and I believe that we often do not fully explore the underlying basic botany of the plants to reveal how they thrive in the field. Presently, EPA is supporting a program on musk thistle control. For that program it was apparent that we did not have sufficient knowledge about the botany of the plant. We hope that such information will reveal weaknesses in the life cycle which can be exploited to permit better control.

URBAN IPM

Let us shift now from the rural to the urban IPM scene. I believe that urban IPM is here to stay. The phrase "urban IPM" has only recently come into use but for some scientists, it's a familiar field because they have been working in it for a long time. I divided this topic into two areas: inside and outside the home. Within the home one finds the control of cockroaches as a major focus. But controls can cover just about anything that crawls into or lives within the home.

The homeowner often gets upset about insect problems. If he can rid his residence of insects, he may often tolerate insects outside his garden. On the other hand, there are many people who do not want to see a single aphid anyplace in their garden. Furthermore, they will use an immense amount of chemicals to make their garden insect free. Therefore, I think we are going to need more work on control of insect pests in lawns and gardens. One can extend this topic to other urban situations such as publicly maintained street right-of-ways and parks

where tailor-made urban IPM practices will be useful. It is unfortunate that the number of scientists working in this area is small.
Progress will be slow.

FUTURE RESEARCH NEEDS

Let us consider some other future IPM frontiers. While predictions
are always lacking in accuracy, I would say the following are likely
to occur. I think that the ability of man to culture large numbers of
desirable predators is going to be an area of fruitful research. It
would be highly desirable if we could have the ability to readily
utilize inundative predatory insect releases under certain situations
to control damaging insect outbreaks. This might reduce the need to
establish predator populations which must survive the winter or some
other environmental stress. Would it not be helpful if we could just
order, as one of the speakers said, five gallons of some control insect;
release them and know that they would survive long enough to take care
of a particularly bad situation?

I think that there will be a continuing effort to find and utilize
fungal, bacterial, and viral pathogens for insect control and especially
for biotic control of weeds. Regretfully, this is another area which
does not have enough people working in it. That does not, nevertheless,
reduce the need.

Now I wish to discuss something which I think is a very basic approach
to insect control and which will perhaps receive the most attention in
the future. I refer to biochemical regulation of our pest populations.
That is, the use of various insect and plant growth regulators to
control unwanted pests. In order to utilize these techniques we need
to have a greater physiological understanding of the control mechanisms
of undesirable pest insects and weeds. If one considers that all living
entities are composed of chemical compounds, it follows that one should
eventually be able to unravel the mystery enough to attain control.
This is not an easily accomplished goal. Some of the molecules, information molecules, are transient and fragile; they come into being,
react, and are gone, leaving almost no way of identifying them. Nevertheless, the insects respond to them. The insect can tell when they are
on an alfalfa plant; they can discriminate types of food; they can
disciminate hostile and acceptable environments where they can live;
they can identify and find mates for reproduction. These and other responses are chemically controlled within the physiology of the pest. If
we consider what we know versus what is obviously taking place in nature,
it's not difficult to deduce that there remains a great deal we do not
yet understand.

It has been pointed out that we know very little about the chemistry that
controls plant resistance to insects. Sometimes the mechanisms are
quite unusual. We heard today at this meeting that in one situation
varying plant susceptibility to insect attack was simply due to greater

or lesser amounts of silicon in the plant cells. In those plants with high silicon, the mandibles of the insects were worn down causing the insects to starve. This is an amazing revelation. How many plant breeders have considered a pest control mechanism involving an increase in the silicon content of the plant cell in order to protect the crop? If we knew in greater detail what was needed it might drastically change how plant breeding programs are designed.

Consider for instance how weeds compete and invade into new areas. It may be their growth habit, or it just may be a chemical predisposition orchestrated by the presence of the plant.

We really do not sufficiently understand many things about ecological succession. If we knew more clearly the events taking place we might be able to use that information in agricultural ecosystems. In fact if we used a holistic approach such as ecologists often use we may very well be farther ahead than with our fragmented approach now being used.

My last observation builds somewhat upon all my previous ones. Basically a case has been made for the need for more information about pests and how they function. How the data (information) should be handled is the last point I wish to discuss. Succinctly stated, I believe the most efficient and long lasting way to use the data is to incorporate it, whenever possible, into mathmatical models which can be revised and improved as more facts are revealed. Models will survive long after the investigator has moved to other interests. Subsequent additions by others will make such models good cumulative distillations of the important aspects of pest organisms, their responses to each other and to other biological entities and to the various environments in which they live. With the availability of these increasingly accurate mathematical models, the control of pests will become more attainable because we will be able to readily discern the achievable from the unachievable.

AGENDA

THURSDAY, 30 MARCH

8:30 am Registration

9:15 am Introduction

Elizabeth Wright Ingraham, President,
Wright-Ingraham Institute

James Lehr, Environmental Protection Agency
Region VIII

Introduction to morning speakers: Robert Simpson,
Colorado State University

9:30 am
CURRENT PRACTICES IN INSECT PEST CONTROL
David Pimentel, Cornell University

10:20 am
BIOLOGICAL CONTROL BY USE OF NATURAL
ENEMIES
Robert van den Bosch, University of California

11:10 am
CULTURAL METHODS FOR PEST CONTROL
Theo F. Watson, University of Arizona

12:00 Lunch

1:00 pm
Working Sessions: PEST PROBLEMS FACING THE
REGION

CROP PESTS (Rm. C)
Eugene Heikes, Extension Professor, Colorado
State University
William Hantsbarger, Extension Professor,
Colorado State University
FOREST PEST MANAGEMENT (Hort. Hall)
Robert Stevens, Rocky Mtn. Forest & Range Exp.
Stat., USFS
Kenneth Lister, Forest Pest Management, USFS
Dave Leatherman, Colorado State Forest Service
LIVESTOCK AND RANGE PESTS (Rm. B)
Austin Haws, Utah State University
Lowell McEwen, Fish & Wildlife Service
URBAN AND HORTICULTURE PESTS (Rm. A)
Byron Reid, Pest Control Association, Regional
Chapter
John Quist, Colorado State University

3:15 pm
Plenary Session: summary of working sessions

4:15 pm
EPA AND PEST MANAGEMENT Charles Rees
Office of Pesticide Programs, Environmental
Protection Agency, Washington, D.C.

4:50 pm   Adjourn

FRIDAY, 31 MARCH

Introduction to Morning Speakers, Beatrice Willard,
Colorado School of Mines

9:00 am
USDA AND PEST MANAGEMENT, Richard L.
Ridgway, Staff Scientist, Science Education
Administration, Beltsville, Maryland

9:35 am
ECONOMICS OF PEST MANAGEMENT, Raymond
Frisbie, Cooperative Extension Service, Texas A&M

10:10 am
Panel Discussion: IMPLEMENTATION OF
INTEGRATED PEST MANAGEMENT
PROGRAMS. Panel Leader: Leon Moore,
Cooperative Extension Service, University of
Arizona
William Olkowski, Univ. of California
Mesa County Peach Administrative Committee,
Wayne Bain; Allan Jones; Palisade, Colorado
Earlie Thomas, Field & Lab, Inc., Ft. Collins

12:00 Lunch

1:00 pm
Panel Discussion: VIEWS ON PEST MANAGEMENT

Panel Leader: F. Martin Brown, Wright-Ingraham
Institute

Thomas Lasater, Rancher, Matheson, Colorado
Glen Murray, Farmer, Brighton, Colorado
William Tweedy, Cebi-Geigy, Inc., North Carolina
Pauline Plaza, Audubon Society, Lakewood,
Colorado

2:30 pm
Working Sessions: CASE STUDIES ON
INTEGRATED PEST MANAGEMENT
PROGRAMS
ALFALFA, Donald W. Davis, Utah State Uni-
versity
Robert Simpson, Colorado State University
URBAN PEST MANAGEMENT, William and Helga
Olkowski, University of California and John Muir
Institute
BREEDING INSECT RESISTANCE IN PLANTS:
WHEAT AND HESSIAN FLY, Robert Gallun.
Science Education Administration (USDA)
Purdue University
THIRD GENERATION PESTICIDES: PHERO-
MONES AND HORMONES, E. Mitchell, Science
Education Administration, Gainsville, Florida

4:00 pm
Plenary Session: CURRENT AND FUTURE RE-
SEARCH NEEDS, Kenneth Hood, Environmental
Protection Agency, Research & Development,
Washington, D.C.

4:40 pm   Adjourn

CONFERENCE PARTICIPANTS

DR. DAVID AKEY, USDA, FR-SEA, Arthropod-borne Animal Disease Research Lab, Denver, Colorado

JANET ALBRIGHT, Colorado State University, Ft. Collins, Colorado

DEBRA ALLEN, student, Colorado State University, Ft. Collins, Colorado

ROBERT ANDERSON, Denver Housing Authority, Denver, Colorado

WAYNE BAIN, Executive Secretary, Mesa County Peach Administrative Committee, Palisade, Colorado

CAROL BARBER, Aurora Vo-Tech, Aurora, Colorado

DR. A. H. BAUMHOVER, USDA, FR-SEA, Tobacco Research Lab, Oxford, North Carolina

DR. MICHEAL BREED, EPO Biology, University of Colorado, Boulder, Colorado

FRANCO BERNARKI, Manager, Superior Farming Company, Tucson, Arizona

F. MARTIN BROWN, Staff Naturalist, Wright-Ingraham Institute, Colorado Springs, Colorado

HERB CHILDRESS, Master Gardener, Colorado Springs, Colorado

WAYNE COLBERG, Cooperative Extension Service, North Dakota State University, Fargo, North Dakota

PHYLLIS CORCHARY, Cooperative Extension Service, Jefferson County, Colorado

DR. DONALD DAVIS, Utah State University, Logan, Utah

PIETER DE JONG, Administrative Staff, Wright-Ingraham Institute, Colorado Springs, Colorado

STEVE DENNIS, Stearns-Rogers Inc., Aurora, Colorado

LAWRENCE R. DE WEESE, U.S. Fish & Wildlife Service, Fort Collins, Colorado

CAROLINE DE WILDE, Cooperative Extension Service, El Paso County, Colorado Springs, Colorado

DOROTHY DICKERSON, Horticultural Advisory Council, Colorado Springs, Colorado

DENNIS DOWNING, teacher, Aurora Vo-Tech, Aurora, Colorado

DR. KENNETH DOXTADER, Horticulture Dept., Colorado State University, Ft. Collins, Colorado

LESLIE EKLUND, IPM consultant, Western Field Technology, Palisade, Colorado

DR. H. E. EVANS, Dept. Zoology and Entomology, Colorado State University Ft. Collins, Colorado

DOROTHY FALKENBERG, Cooperative Extension Service, Golden, Colorado

CATHRYN FLANAGAN, student, Colorado State University, Ft. Collins, Colorado

KENNETH FORDYCE, Denver Housing Authority, Denver, Colorado

CAROLE FORSYTH, Denver Audubon Society, Northglenn, Colorado

J. H. FOWLER, Chairman, Biocides Recyling Committee, Enos Mills Group of the Sierra Club, Denver, Colorado

DR. RAYMOND FRISBIE, Texas Cooperative Extension Service, Texas A&M, College Station, Texas

DR. ROBERT L. GALLUN, USDA, FR-SEA, Purdue University, West Lafayette Indiana

KEITH E. GOOSMAN, teacher, Poudre School District, Ft. Collins, Colorado

W. L. GORDON, Colorado Agricultural Chemicals Association, Denver, Colorado

LYNNE GRACE, Aiken Audubon Society, Colorado Springs, Colorado

DR. WILLIAM HANTSBARGER, Extension Professor, Department of Zoology and
Entomology, Colorado State University, Ft. Collins, Colorado
RICHARD HART, Northwest Missouri State University, Maryville, Missouri
DR. AUSTIN HAWS, Biology Department, Utah State University, Logan, Utah
EUGENE HEIKES, Extension Professor, Weed Research Lab, Colorado State
University, Ft. Collins, Colorado
DR. KENNETH HOOD, Office of Research and Development, Environmental
Protection Agency, Washington, D.C.
CAROLYN HUISJEN, LCTC Horticulture Society, Ft. Collins, Colorado
BARBARA HYDE, Cooperative Extension Service, Boulder County, Colorado
ELIZABETH WRIGHT INGRAHAM, President, Wright-Ingraham Institute, Colorado
Springs, Colorado
DR. ROBERT H. JONES, Research Entomologist, USDA, FR-SEA, Denver,
Colorado
DR. JAY B. KARREN, Extension Entomologist, Utah State University,
Logan, Utah
LEWIS KEENAN, USDA, APHIS, Denver, Colorado
RICHARD KEIGLEY, National Park Service, Denver, Colorado
EDWARD KEITH, Biology Department, University of California at Santa Cruz,
Santa Cruz, California
JAMES KEITH, U.S. Fish & Wildlife Service, Patuxent Wildlife Research
Center, Denver, Colorado
DAVID N. KIMBALL, Denver, Colorado
THOMAS LASATER, The Lasater Ranch, Matheson, Colorado
MARIE LAUFER, student, University of Colorado, Colorado Springs, Colorado
DAVID LEATHERMAN, Colorado State Forest Service, Ft. Collins, Colorado
JAMES LEHR, Hazardous Waste Division, Environmental Protection Agency,
Region VIII, Denver, Colorado
KENDALL LISTER, Forest Pest Management, Rocky Mountain Region, U.S Forest
Service, Denver, Colorado
JEANNE MALONEY, Horticultural Advisory Council, Colorado Springs,
Colorado
JOHN B. MC CLAVE, Cooperative Extension Service, Summit County, Frisco,
Colorado
LOWELL C. MCEWEN, U.S. Fish & Wildlife Service, Patuxent Wildlife
Research Center, Ft. Collins, Colorado
DALLAS MILLER, Pesticide Branch, Environmental Protection Agency,
Region VIII, Denver, Colorado
DR. J. MINTON, EPO Biology, University of Colorado, Boulder, Colorado
DR. EVERETT R. MITCHELL, USDA, FR-SEA, Insect Attractants Laboratory,
Gainesville, Florida
DR. LEON MOORE, Cooperative Extension Service, University of Arizona,
Tucson, Arizona
RONALD MORROW, City Park and Recreation Department, Colorado Springs,
Colorado
DONALD NELSON, Boulder, Colorado
EUGENE NELSON, Cooperative Extension Service, Alamosa County, Alamosa,
Colorado
LYNDA M. NIELSEN, Loveland, Colorado
DR. WILLIAM OLKOWSKI, Center for the Integration of the Applied Sciences,
John Muir Institute, Berkeley, California

HELGA OLKOWSKI, Center for the Integration of the Applied Sciences, John Muir Institute, Berkeley, California
TIM ORTNER, Representative for President of the Colorado State Senate, FRED ANDERSON, Denver, Colorado
ANDREW PIERCE, Superintendent of Conservatory, Denver Botanic Gardens, Denver, Colorado
DR. DAVID PIMENTEL, Department of Entomology, Cornell University, Ithaca, New York
PAULINE PLAZA, Western Environmental Science Program, National Audubon Society, Lakewood, Colorado
JOHN POHLY, Larimer County Vo-Tech, Fort Collins, Colorado
DR. JOHN A. QUIST, Department of Zoology and Entomology, Colorado State University, Ft. Collins, Colorado
CHARLES REESE, Office of Pesticide Programs, Environmental Protection Agency, Washington, D.C.
STUART REEVE, Ft. Collins, Colorado
BYRON REID, Regional Vice President, Pest Control Association, Colorado Springs, Colorado
RICHARD RIDGWAY, National Program Staff, Federal Research, Science and Education Administration, USDA, Beltsville, Maryland
KEVIN ROSENFOFFER, student, Colorado State University, Ft. Collins, Colorado
ALEX SCHUETTENBERG, Ft. Collins, Colorado
J. F. SHAUGHNESSY, Monte Vista, Colorado
DANIEL SHEEHY, Stillpoint Hermitage, Manitou Springs, Colorado
FRANK SIEBURTH, Castle Rock, Colorado
MARGRET SIKES, Denver Botanic Gardens, Denver, Colorado
JOAN E. SIKKENS, Denver Audubon Society, Aurora, Colorado
DR. ROBERT SIMPSON, Department of Zoology and Entomology, Colorado State University, Ft. Collins, Colorado
RONALD STEE, South Dakota Department of Agriculture, Pierre, South Dakota
BARBARA STEINMEYER, Department of Parks & Recreation, City of Westminster, Westminster, Colorado
DR. ROBERT STEVENS, Rocky Mountain Forest and Range Experiment Station, U.S. Forest Service, USDA, Ft. Collins, Colorado
ADAIR STONER, USDA, FR-SEA, Honey Bee, Pesticides & Diseases Research, University Station, Laramie, Wyoming
CURT SWIFT, Cooperative Extension Service, El Paso County, Colorado Springs Colorado
JERROLD SWITZER, Department of Parks & Recreation, Colorado Springs, Colorado
EARLIE THOMAS, President, Field & Lab, Inc., Ft. Collins, Colorado
DR. ROBERT VAN DEN BOSCH, Division of Biological Control, University of California at Berkeley, Berkeley, California
MARK WALMSLEY, Dept. of Zoology and Entomology, Colorado State University, Ft. Collins, Colorado
JUDY WARD, Denver Audubon Society, Denver, Colorado
DR. THEO WATSON, Department of Entomology, University of Arizona, Tucson, Arizona
LOIS WEBSTER, Aurora, Colorado
WAYNE WEHLING, Arvada, Colorado
DR. BEATRICE WILLARD, Environmental Sciences, Colorado School of Mines, Golden, Colorado

BRUCE WILLIAMS, Superior Farming Company, Tucson, Arizona
PANDORA WILSON, Master Gardener, Jefferson County Extension, Lakewood, Colorado
MEREL O. WOODS, Cooperative Extension Service, Arapahoe County, Littleton, Colorado
DAVE WOODWARD, Broadmoor Greenhouse, Colorado Springs, Colorado
MICHEAL WYBLE, Department of Parks & Recreation, City of Westminster, Colorado

INDEX

A

Acryrthosiphum pisum, 3, 63
agrichemical industry, view on IPM, 94
agroecosystem, 93, 100
    diversification of, 60
AID (Agency for International
Development),
    role in IPM, 112
alfalfa,
    ecosystem model for, 41
    harvesting dates, 62
    IPM in, 40-42
    pests, 42
    resistant varieties, 3, 40
    strip-cutting of, 62
aphid,
    linden, 123
    pea, 3, 63
    spotted alfalfa, 65
Aphidus smithi, 38
atmospheric permeation, 71

B

Bacillus thuringiensis, 22, 124
beetle,
    ambrosia, 76
    mountain pine bark, 25-27
biological control,
    classic introduction programs, 7, 52,
      124
    definition of, 49
    host specificity of control agent,
      52-53, 125
    in alfalfa, 37-39
    in peach orchards, 37
    naturally occurring, 54-55, 124
    of klamath weed, 38
    regional programs, 37-39
    weeds, 52-54
biotype, 83
Blissus leucopterus, 3
bollworm, pink, 66-69, 71-72
    additive cultural practices on, 65-67
budworm, eastern spruce, 75

C

CEQ (Council on Environmental Quality),
    112
chemical sex attreactants, 8-10
chinch bug, 3
codling moth, 74
Colorado Department of Agriculture,
    biological control programs, 36-38
    seed embargoes, 23
Commonwealth Institute of Biological
Control, 53
corn,
    earworm, 73
    insect control in, 22
    leaf blight, southern, 5, 13
    rootworm, 4, 22, 61
cotton,
    control of lygus in, 62-63
    cultural controls in, 4
    early IPM program, 99
    reduction of insecticide use in,
      102-103, 106-107
    scouting program in, 101-102
crop,
    breeding insect resistance in,
      2, 81-85
    effects of reduced genetic diversity
      in, 5
    losses to pests, 1
    losses, postharvest, 1
    insect resistant varieties, 2-3
    rotations, 4, 61
cultural control,
    attributes of, 60
    cropping systems, 4-5
    definition of, 59
    goals of, 60
    in cotton, 66-69
    planting dates, 6-7, 61
    strip cutting systems, 62-63
    trap crops, 64

D

Delaney Clause, 110
Dendrosoter protuberans, 38

139

diapause, 67
Diparopsis castanea, 72-73

E

economics,
   in chemical weed control, 23
   in urban shade tree IPM, 126
   of pest management in cotton,
     105-107
economic thresholds, 23, 93, 105
economic poison, 109 (*see also* pesticide)
ecosystem, effects of pesticides to,
   12-13, 31-32
Endrin, 23
EPA (Environmental Protection Agency),
   regulatory responsibilities for
     pesticides, 111
   role in implementing IPM, 113

F

field bindweed, 23
FIFRA (Federal Insecticide, Fungicide, &
   Rodenticide Act), 109
forest, pest management, 25-27

G

gossyplure, 72 (*see also* pheromone)
government, policies on range management, 43, 45
grasshopper, control in range, 24-25
grower organizations,
   cotton, 101-103
   orchards, 35-36
gypsy moth, 8

H

habitat management, 26, 59
herbicides, (*see also* pesticide)
   effects on insect populations, 13,
   regional use of 23-24
   use of 2, 4-D in weed control, 23
Hessian fly, 3, 81-83
   fly free planting dates, 61-62
   genetic control of, 85
HEW (Health, Education & Welfare),
   role in IPM, 112
host specificity of parasites, 52-53, 125

HUD (Housing & Urban Development),
   role in IPM, 112-113

I

injury levels, 122 (*see also* economic
   threshold)
insecticide (see also pesticide)
   aldrin, 24
   malathion, 24
   reduction in shade trees, 123
   reduction of use in cotton, 102-103,
     106-107
integrated pest control, 91 (see also
   integrated pest management)
integrated pest management (IPM)
   consultants, 36, 124
   definition of, 1, 59, 91
   future research needs, 129-132
   implementation of, 118-119
   in alfalfa, 40-42
   in cotton, 65-69, 101-103, 105-107
   in home gardens, 120, 121
   in orchards, 36
   in pine bark beetle control, 25-27
   research needs in range, 43-45
   urban, 121-124
   urban research needs, 130
Inter-agency Regulatory Liaison Group, 35
interdisciplinary research in range, 44-45

L

Labops sp., 44
labor, in agriculture, 29, 92
Lygus sp., 95
   effects of stripcutting on, 4, 62

M

Macrocentrus ancylivorus, 36-37
Mayetiola destructor, 3, 61-62, 81
mediterranean fruit fly, 8
Miller Admendment, 110
mite,
   Bank's grass, 22
   two spotted, 37-38
modelling,
   crop management systems, 132
   in alfalfa, 91-92
multicrop approach, 102

N

National Academy of Sciences (NAS), 113
National Agricultural Chemicals
 Association (NACA), 92, 94
NSF (National Science Foundation),
 support of IPM, 113
Neosema locustae, 23-24

O

olfactory attractants, 7-10 (see also
 semiochemicals, pheromones)
organophosphates, 99-100

P

parasite, definition of, 49
parasitoid, definition of, 49
pathogen, definition of, 49
pesticides,
    benefits, 9-11, 91-92
    carbamates, 99
    development of resistance in insects,
     100
    distribution of use, 10
    effects on crop physiology, 13
    effects on non-target species, 23
    effects on public health, 12
    effects on raptor populations, 32
    environmental costs, 11-12, 31-33
    residues in produce, 12
pest management (see also IPM),
    basic elements of, 100-101
    components of, 99
    definition of, 1
    economics of, 105-107
    in forests, 25-27
    of cotton, 101-103
    systems, 93
    use of pesticides in, 94
    use of resistant varieties in, 81
pheromones,
    atmospheric permeation, 71
    dosage response, 10
    for population sampling, 75-76
    potential in pest management, 7-10,
     71-74
    sex attractants, 71-73
    use in stored products, 74
Porthetria dispar, 8
predator, definition of, 49

R

redbanded leaf-roller, 8
rootworm, *see* corn, rootworm

S

saltmarsh caterpillar, 63-64
sand wireworm, 64
semio-chemicals, 72
sorghum,
    resistant varieties, 3
    cultural controls in, 62
soybeans, 4

T

technology transfer,
    rural to urban, 122-124
    to home gardens, 120
Texas Cotton Pest Management Program,
 107
thistle,
    bull, 23
    Scott's, 23
    musk, 23, 38, 130
trap crops, 64
Trioxys curuicandus, 23
tussock moth, 56, 75
Typhloromus occidentalis, 37

U

USDA,
    early IPM projects, 82, 117-118
    mission, 177
    role in implementing IPM, 118-119

W

water management, 65
weed,
    biological control of, 52-54
    contemporary control strategies,
     23-24
    control districts, 24
weevil, alfalfa, 40-42 62 (*see also* alfalfa)
wheat, 23
    use of resistant varieties, 82-85

141

Printed and bound by CPI Group (UK) Ltd, Croydon, CR0 4YY

23/10/2024

01778244-0001